A GUIDE TO THE PHOTOVOLTAIC REVOLUTION

A GUIDE TO THE

PHOTOVOLTAIC REVOLUTION

Paul D. Maycock & Edward N. Stirewalt

Rodale Press, Emmaus, Pa.

To my wife
the Reverend Roma L. Maycock
—P.M.

To my wife
Marcia Winton Stirewalt
superb teacher
—E.S.

Printed in the United States of America on recycled paper containing a high percentage of de-inked fiber.

Illustrations: Michael Prendergast and Frank Rohrbach

Library of Congress Cataloging in Publication Data

Maycock, Paul D.
A guide to the photovoltaic revolution.

Rev. ed. of: Photovoltaics, sunlight to electricity in one step. © 1981
Includes bibliographies and index.
1. Photovoltaic power generation. I. Stirewalt, Edward N. II. Title.
TK2960.M39 1985 621.31' 244 83-15413
ISBN 0-87857-532-4 hardcover
ISBN 0-87857-534-0 paperback

2 4 6 8 10 9 7 5 3 1 hardcover
2 4 6 8 10 9 7 5 3 1 paperback

CONTENTS

LIST OF APPENDIXES

PREFACE

Today more than ever, solar photovoltaic (pronounced fō'-tō-vōl-tā'-ĭk) energy promises to become a leading source of electrical power, ultimately rivaling the largest existing sources of electricity.

The onset of this revolution in energy affairs has been exceedingly quiet, considering its potential. For more than two decades, the work has gone steadily forward in the laboratories and testing centers of universities, government research centers, and industry. Reliable products are now on the market at prices that make photovoltaics, or PV, economically feasible in many situations.

As prices continue their decline and as photovoltaic systems improve in quality, reliability, and the capacity to convert sunlight directly into electricity, the "revolution" is now entering a new phase. Changes have been coming at a rapid pace. For example, to the surprise of many in the field, electric utilities have become the largest purchasers of photovoltaic electricity. At the same time, one technological approach thought to be an unlikely candidate for widespread use has suddenly emerged as a leading contender. Another significant development is that private industry is now spending twice as much as the government on research and development for photovoltaics.

Changes such as these and rapid motion in all parts of the photovoltaics field have led us to revise our book, *Photovoltaics—Sunlight to Electricity in One Step,* published in 1981. The present volume is an extensively updated and expanded edition of that work.

It is a challenge to cover in a volume of modest size a subject of such technical sophistication and diversification as photovoltaic power. PV systems are simple looking and easy to use, yet they are marvels of high technology. In addition, the range of uses, the apparently vast opportunity for growth, and the far-reaching social implications of widespread photovoltaic use add to the excitement of what is taking place. The annual market for photovoltaic systems may well reach or exceed $20 billion globally by the turn of the century.

We have attempted to provide an overview that will be helpful to all who may be seeking a general introduction to the field—students, scientists and engineers, homeowners, business persons, investors, tradespersons, potential commercial and industrial users, those interested in protecting and improving the environment, and persons everywhere who hope that from the vast storehouse of technology will come more things that are truly beneficial to humankind.

Photovoltaics is already helping many people in many different ways. The feasibility of PV as a power source has been clearly demonstrated in the field, and its ultimate potential has been dramatically outlined for all the world to see. It is clear that a revolution of sorts is in the making. Its dimensions and timing, not to mention its consequences, can only be guessed at, but the prospects are fascinating.

The challenge now, to which we are deeply committed, is to bring the photovoltaic revolution one step nearer. It is a worthy cause in which everyone who wishes can find a way to participate.

We renew our thanks to all who helped make the first book a success. In addition, we are grateful to Dr. Lloyd Herwig of the U.S. Department of Energy (DOE), who was helpful with statistics regarding international developments, and to Paul Brown, DOE's electric vehicle technology director, who was kind enough to review the section on electric vehicles. Jay Holmes of DOE's renewable energy staff was helpful in several particulars.

We extend our appreciation to Eloise Bean of Planning Research Corporation for her thoughtfulness and support in connection with this as well as the first volume.

Finally, to our wives go deepest gratitude for their continued patience, understanding, and forbearance in this latest effort.

All who have helped, however, shall remain blameless. The authors alone bear full responsibility for the result.

P.D.M.
E.N.S.

Fairfax County, Virginia
July 1984

CHAPTER 1
WHY WE NEED PHOTOVOLTAICS

Photovoltaic devices—the solar cells which convert sunshine directly into electric current—first came to the attention of the general public with the advent of America's space program. Like the vehicles which carried them, photovoltaic cells were seen as exotic, impractical, and expensive. The solar cells that accompanied the Vanguard I were painstakingly made by hand, and cost more than $1,000 per Watt. Such a cost would be clearly prohibitive for a typical household, which needs 5,000–10,000 Watts to satisfy its peak electrical demand.

In nearly a century and a half since the photovoltaic effect was first observed in 1839, the technology has grown from a laboratory curiosity into a mature science. With the myriad practical uses found for it in the last 20 years, photovoltaics is on the brink of exploding into the energy marketplace.

Originally hand-built and labor-intensive, solar cells can now be made on an assembly-line basis. Originally designed for the most specialized of uses, they can now be found on self-powered wristwatches, navigation buoys, and buildings. Prohibitively expensive at first, they are now at most only a few years from being directly competitive, on a cost-per-kilowatt-hour basis, with electricity generated by traditional water-, coal-, oil-, or nuclear-driven plants.

Photovoltaic power, contrary to the common perception, is no longer an exotic technology. To give some idea of just how far the reality of photovoltaics departs from the popular misconception, here are a few predictions:

1. Photovoltaics will be fully economic for massive private use before a major utility can design, purchase, and install its next new large-scale power plant.
2. Photovoltaic systems installed on the roofs of residences in the United States will be fully competitive with other sources of electricity within the next ten years.

Figure 1-1 Daytime WBNO, a radio station broadcasting from Bryan, Ohio, draws most of its power from a 15-peak-kilowatt photovoltaic array. A battery is used to store excess energy. (Courtesy: U.S. Dept. of Energy)

3. If we seriously begin to adopt photovoltaics now, as much as 10 percent of the nation's electric energy can come from this source by the year 2000. And after that, the percentage should increase steadily.

Photovoltaics offers the surest way out of the growing energy crisis in which we are mired. We must treat this technology as a practical reality to be integrated both into decisions about energy policy and supplies, and into the technical and political assumptions that underlie our economy and sustain our society.

In a very real sense, and in whatever form, energy is the lifeblood of this society. Whatever the source—oil, coal, hydro, nuclear, gas—the use of energy is all-pervasive. Energy lights and heats our homes, offices, and factories. It powers the machines of industry and transportation. The clothes we wear; the food we eat; the buildings where we live and work; the goods we produce, such as dye, lubricants, metals, glass, plastic; and the services we supply, such as transportation and communication—energy is inextricably woven into it all.

There will be no better time than now to face the reality that our store of fossil fuels, our hydro sites, even our supplies of uranium are finite. The OPEC nations, unwittingly and for reasons other than altruism, have done the world a favor by bringing to our attention, in stark and unambiguous terms, this central fact.

We in the United States are one-twentieth of the world's people, yet we consume one-third of its energy. Ignoring questions of equity for the moment, the interruption of energy supply is more disruptive to a highly industrialized society such as ours than to a less-developed one. Since the first oil embargo of 1973, we have recognized our vulnerability. But in spite of hand wringing and lip service, we have made little genuine progress toward resolving our energy dilemma.

In the absence of a focused national commitment, the real and perceived energy shortages of the 1970s exacerbated tensions between different elements of our society. The widespread conviction that we were being taken for a ride by "big oil," and the bumper stickers that read "Drive Eighty and Freeze a Yankee" or "If you like our Post Office, you'll love nationalized oil," were symptomatic. Even with the reduced oil imports of the 1980s, a continuing consequence of our energy situation is the chronic and massive balance-of-trade deficit. It is small comfort that other industrialized nations lacking substantial domestic supplies are economically hurt more than we are.

This trade deficit aggravated the inflationary spiral begun during the Vietnam war years. At the same time that rising oil prices increased the costs of heating, manufacture, and transportation, the billions of dollars we have sent the OPEC nations have exerted a devaluating influence on our currency. An editorial in the *New York Times* pointed out that "energy is still the problem—inflation is merely the mirror…no one is going to solve the inflation problem without first solving the energy problem."[1] Even with the subsequent easing of the energy crisis in the 1980s, Energy Secretary Paul Hodel said that we must have a balanced program based on conservation coupled with conventional and renewable energy sources. It is obvious that something must be done to stanch the economic hemorrhage caused by the nation's appetite for energy.

On the one hand, we long for ultimate solutions to the energy problem. The breeder reactor, some argue, would save us if only environmentalists and kindred malcontents would get out of the way. The fusion generator will provide us with cheap energy, but it may require that we invest billions and wait 50 years. In another scheme, orbiting photovoltaic satellites could beam energy to earth by microwave where it would be reconverted to electricity in quantities to meet the total U.S. demand.

On the other hand, such solutions as synfuels, more nuclear plants, coal-to-methane conversion, and others are at root a recommendation for the "more of the same" attitudes that got us in trouble in the first place. They commit us even further to the existing high-capital, central-distribution model that involves increasingly unacceptable external costs: pollution, containment of radioactive wastes, supply monopoly, operating and maintenance costs, and inelastic fuel supply.

"More of the same," in addition to incurring these external costs, may create new problems. For example, in the summer of 1979, Congress and President Carter were pressing ahead with a massive synthetic fuels program, which involved mining and processing vast reserves of coal and oil-rich shale in several western states to produce a number of synthetic fuels (gas, liquid fuels, and refined coal, among others). However, processing these deposits requires large amounts of water, a precious natural resource perennially in short supply there. Moreover, the economic feasibility of the program has been questioned by many critics. By 1985, the synfuels program had completed no projects and was virtually dead.

We live in a highly technological society, and an inextricable element of this is our massive need for energy. This need is neither good nor bad in and of itself. We have devoted enormous amounts of capital to provide for our energy needs, and patterns of decision-making about energy have been institutionalized to the point that it is now very difficult to think about energy in other than "more of the same" terms.

This "more of the same" attitude tries to treat the energy crisis as a supply problem—solvable with more mines, more wells, and more reactors—when in reality it has become a structural problem. For the most part, "more of the same" is an extrapolation of the energy supply strategies we have followed since centrally produced electric power first became a reality. Considering the number of people, physical plants, and equipment already committed to this structure, such a response is understandable, but it is wrong-headed.

There is no need to rehearse here the shortcomings of the centralized fossil fuel and nuclear model to which our country and all industrialized nations are presently committed. By the same token, there is no need to do a cost-benefit analysis of the indirect methods of tapping the sun: wind power, driven by the sun's heating the earth and its atmosphere; ocean thermal, which exploits the sea's temperature differentials; biomass, which binds the sun's energy in chemical bonds in plants. They should be explored; diversity of sources, rather than some single monolithic solution, is the best strategy. But this book is not a global discussion of energy policy. It is about photovoltaics and how this technology not only can help us out of our present plight, but also substantially change our world for the better.

We restrict ourselves when we think of solar power only as low-grade energy available as heat, which buildings can capture and distribute "passively" through proper location, orientation, and design, or which collectors can

Figure 1-2 The photovoltaic-powered RAMOS weather station at the edge of New Mexico's vast eastern plains reports weather data hourly via satellite to National Weather Service offices. (Courtesy: NASA Lewis Research Center)

Figure 1-3 These products manufactured by Sanyo use amorphous silicon photovoltaic cells. (Courtesy: Sanyo Electric, Inc.)

capture and distribute "actively" with circulating liquids or forced-air systems. Several proposals for collecting sunlight, a very diffuse source of energy, involve concentrating its rays so that the resultant heat performs work such as generating steam, which in turn drives turbines to produce electricity. But these systems commit us to the fixed, capital-intensive, cumbersome, complicated central-distribution model that is a part of our current energy problems.

In contrast, photovoltaic technology is both simpler and more sophisticated than either the traditional fuel-driven or the proposed solar thermal-electric plants. Rather than elaborate and costly machinery—all too subject to Murphy's Law—photovoltaics uses the fundamental properties of matter itself to produce power quietly, safely, and with virtually no operation and maintenance costs. Because photovoltaic systems are modular and can be produced in size and type to fit almost any need, they permit local, decentralized installation at the point of need, and can, in very simple fashion, be designed to match the output to the user's purposes.

The best analysis of our energy situation appears in the Harvard Business School study *Energy Future.*[2] With deletions, but without distortion, the authors' conclusions can be boiled down to this:

> The purpose of U.S. energy policy should be the managing of a *transition* from a world of cheap imported oil to a more balanced system of energy sources. . . . The nation has only two major alternatives for the

> rest of this century—to import more oil or to accelerate the development of conservation and solar energy. . . . Because environmental and health problems have increasingly become issues in the political process, the outlook for coal is considerably less promising than officially projected. . . . Nuclear energy has a set of external costs even more controversial [than coal]. . . . At most, during the next dozen years nuclear energy might add a million and a half barrels a day of oil equivalent thus reaching three million barrels per day. This level would require the continued operation of *all* existing nuclear capacity plus the completion and operation of all new capacity currently under construction or on order. That we stress is a very bullish scenario. . . . The cornerstone of our thinking is that conservation and solar energy should be given a fair chance in the market system to compete with oil and other traditional sources . . . a more significant change could occur from the widespread use of on-site solar technologies.

What then is the problem? If photovoltaics is safe, clean, and reliable, and if there has been a substantial reappraisal of the role of energy in our society, why aren't photovoltaic systems out in the marketplace where the Harvard Business School says they should be?

The answer is that photovoltaic systems are out there for some uses, in some places. Part of the problem is that people are slow to give up outdated ideas about photovoltaics that match the reality of 20 years ago, not of today. One such idea, which happens to be wrong, is that photovoltaic cells are physically and ecomonically impractical to produce.

Part of the problem is that people generally don't know very much about photovoltaics. Briefly, photovoltaic devices—solar cells—absorb sunlight and convert it directly into electricity. This is called the photovoltaic effect; "photo" for light, "voltaic" from the name of the Italian scientist, Volta, for whom the Volt is named. When light strikes certain materials called semiconductors, internal voltages are created. The basic scientific principles which underlie this effect are well understood. The solar energy available on a bright summer day at noon is about 1,000 Watts (1 kilowatt) per square meter.[3] This amount of solar energy falling on a single square meter, if converted perfectly to electricity, could power ten 100-Watt bulbs, or two furnace motors, or several 25-inch television sets.

It is not, of course, high noon and summer all the time, and photovoltaic cells do not work at 100 percent efficiency. They presently work at about 15 percent efficiency, with the best performance being about 25 percent. And some areas of the country receive more light than others. But, the same

Figure 1-4 Photovoltaic solar energy house. The Carlisle House near Boston, Massachusetts, completed in Spring 1981, was specifically designed for a 7.5-peak-kilowatt photovoltaic array. The house also incorporates passive solar design and a solar thermal hot water heater. Above: Installing photovoltaic panels on 45° sloping roof. Below: Architect's rendering of completed residence. Steven J. Strong, Solar Design Associates, architect. (Courtesy: Solar Design Associates, Architects and Engineers, Lincoln, Mass.; module installation photo by MIT)

misconception that has hampered the development of passive solar design, the assumption that it will only work in such locations as the Sunbelt or at certain times of the year, has dogged photovoltaics.

It is true that in a year Boston receives only half as much sunlight as Phoenix. But the local cost of electricity is a variable of great potential significance for determining the suitability of a photovoltaic installation. While sunshine varies by a factor of two in the United States, electricity rates vary by a factor of more than six. In 1984, the Boston homeowner paid $0.10 per kilowatt hour; his Phoenix counterpart paid $0.08. In New York City, where electricity cost $0.17 at that time, the potential market for photovoltaics may in fact be better than in Phoenix, even though New York receives less sunshine. When utility companies in New York raise their rates to equal or exceed the cost of electricity produced by photovoltaic systems—an event that may not be too far in the future—New Yorkers will begin buying photovoltaic systems in large numbers.

Another part of the problem is that in our highly industrialized society it is impossible to change course suddenly and very difficult to do so even gradually. There is no deliberate conspiracy to keep photovoltaics out of the marketplace. But there is a lack of commitment to it, due in part to the popular misconception that photovoltaics has little to offer in this century. Political leadership has done more than pay lip service to photovoltaics but as yet has been unwilling to give it a high place in national priorities.

Any major change in something so basic as the way we produce and allocate increasingly dear energy supplies must have the broadest understanding and support of the people affected by those decisions: ourselves. It is only within the last decade that solar power in general and photovoltaic technology in particular have had a credible and reasonably well-informed constituency. The Solar Lobby, a coalition of groups numbering over 80,000 persons, was formed in 1978 as a result of Sun Day.

Government had previously shown some interest in solar energy. As early as 1972, the National Science Foundation wrote that

> . . . a substantial development program can achieve the necessary technical and economic objectives by the year 2020. Then solar energy could economically provide up to (1) 35% of the total building heating and cooling loads; (2) 30% of the Nation's gaseous fuel; (3) 10% of the liquid fuel and (4) 20% of the electric energy requirement.[4]

Six years later, after unparalleled price increases for imported oil, the President's Council on Environmental Quality estimated that photovoltaics might provide 2-8 quads (one quad equals a quadrillion Btu's) of energy annually by the end of the century.[5]

It should not go unnoticed that under three administrations, the budget for photovoltaics rose steadily from $4 million in 1973 to $151 million in 1980. However, under the Reagan administration it had fallen by 1985 to $59 million. Congress has supported the solar effort through significant special legislation. In 1978, the Solar Photovoltaic Energy Research, Development and Demonstration Act (Public Law 95-590), which sets as a goal the doubling of photovoltaic production annually for ten years and authorizes the expenditure of $1.5 billion over that period, was passed. The law foresees a cumulative installed capacity of 4 gigawatts by 1988, by then at system costs of $1 per Watt in 1978 dollars, $1.50 per Watt in 1984 dollars. (Remember, the solar cells of the sixties cost more than $1,000 per Watt.) Recognizing the potential for change inherent in photovoltaics, the law also establishes a special advisory committee with broad public representation to help assure that the public interests are met.

In June 1979 an interagency Domestic Policy Review committee, called for by President Carter and headed by then Energy Secretary James Schlesinger, set a goal for the year 2000: Solar would contribute 19 quads, out of the 95 required, or 20 percent of the nation's energy supply.[6] The Harvard Business School study also envisions 20 percent of our energy needs being met by solar and explicitly points to photovoltaics:

> One can certainly say that photovoltaics thus far constitutes a success story . . . advances can come quickly, for as one specialist has put it, photovoltaics are "embedded in what has been one of the most fertile environments for innovation in the twentieth century—the semiconductor industry." As has been the case in that industry, increased volume will bring increased experience, improved processes, and reduced costs. Investments in research will produce fundamentally new production techniques that could lead to even lower costs.[7]

The Harvard study urges that federal purchases of photovoltaic equipment be used to drive production costs lower, and even suggests that funds previously earmarked for the Clinch River breeder reactor be used for that purpose.[8] In 1984, Congress terminated the Clinch River project, but unfortunately the savings were not designated for use on photovoltaics.

Over the past several years, there has also occurred in some quarters a fundamental rearrangement of the assumptions and attitudes about the role which energy should play in our society. This reconception appears in various guises, sometimes described as "limits to growth," "small is beautiful," "soft energy paths," or "appropriate technology." The proponents hold in common a vision of a decentralized society in which all energy comes from renewable sources, especially those that are continually replenished by the sun.[9]

As the cost of traditional fuels escalates, it becomes possible for reasonable people to propose energy sources and supply structures that only a few years ago would have been regarded as unrealistic and visionary. Wind power, ocean thermal differentials, and biomass are no longer considered "far out" potential sources of energy. It has become intellectually respectable to side with the late Ernest Schumacher, author of *Small Is Beautiful,* that bigness in institutions, business, industry, or government is not better and may in fact be inimical to many human values and aspirations. Granted, as Daniel Greenberg pointed out in the *Washington Post*, prudent people have been a bit put off by solar energy proposals because of "the vast freight of political and philosophical enthusiasms that some of its friends have loaded onto it."[10] We solar enthusiasts have sometimes gone out of our way to proclaim our disenchantment with some of the ways technology has manifested itself, an attitude which along with its opposite, uncritical technology, too often becomes ideological name-calling.

Recently, however, there has been a growing affirmative consensus that in fostering technical development we should give more care to selecting technical approaches and systems that are better matched to human needs and aspirations. The key word is appropriate. There is nothing intrinsically wrong with a technology that fits the needs of the times, and there is nothing intrinsically wrong with bigness in the right place, helping towards proper goals.

One popular formulation of these sentiments is set forth by Amory Lovins, British physicist, author, and long an advocate of appropriate technology. An ideal energy system, he suggests, should first consist of a set of diverse sources, each doing what it does best. Second, these energy sources must be simple and understandable from the user's point of view. Put a different way, our solutions to the energy problem should be our tools, not our masters. The test is not whether a solution involves high technology, but whether the average person can make practical use of it. In addition, the form of energy considered appropriate should be of the right quality, right for the purpose it serves, and properly scaled to the load. Lovins distinguishes among four kinds of energy:

- Low-grade heat for low-temperature processes
- High-grade heat for processes requiring high temperatures
- Mechanical energy providing torque for mechanical functions
- Electricity for electronic and pure electrical demand

Figure 1-5 The world's first solar-powered gasoline station in West Chicago, Illinois, is a demonstration by the Amoco Oil Company, Solarex Corporation, and Semix, Inc. The 5-kilowatt array and battery bank began operating in July 1980. The day it opened, a violent thunderstorm blacked out power in the city. This was the only station pumping gas that afternoon. (Courtesy: Solarex Corporation)

By carefully choosing technology appropriate to its use, he argues, we can maintain our present standard of living while using half the energy we now consume. For example, electricity, which is high-quality energy, should not be used for heating. Moreover, the by-product heat from electric generating plants should not be wasted but put to practical use. The use of solar thermal energy for heating rates high on Lovins's list of appropriate function matched to energy quality. He has proposed that his "soft path" would eliminate the need to add more conventional electricity generating capacity in the United States. It is unlikely that, as a practical matter, we can avoid constructing at least some new power plants. Still, photovoltaics meets the "soft path"

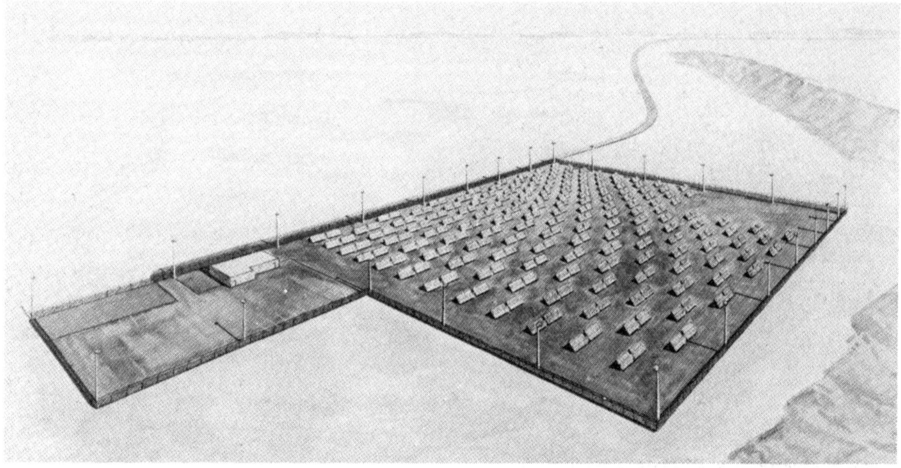

Figure 1-6 Above: In this 350-kilowatt village power project in Saudi Arabia, 2-axis tracking arrays produced by Martin Marietta are used. Below: Closeup of a single array. Sunlight is concentrated on each cell by a plastic Fresnel lens. (Courtesy: Martin Marietta Aerospace)

criteria proposed and is an attractive source of energy in a number of important respects.

In one light, this is pretty heady stuff. But, somehow, there has remained an air of unreality about these activities. The net effect of the projections by blue ribbon committees, academics, and some officials has been to establish in the public mind an impression that renewable energy, and specifically photovoltaics, is meant for the next generation and certainly offers no possible help in the present energy crisis. Many of these studies are shot through with a presumption of helplessness. Like the use of outdated energy supply assumptions and strategies for projections into the future, the idea that photovoltaic technology is only for the next century is simply wrong and needlessly impedes present-day commitments to photovoltaic development.

In another way, simplistic solutions to the energy problem, often presented by some quarters of the solar movement, unfortunately have contributed to the same air of unreality and heightened the public sense of frustration. Solar energy does not mean an across-the-board return to old-fashioned woodstoves and windmills. Far from it. But we do not belittle what the solar enthusiasts have accomplished. Their passion is a reflection of a deeper yearning throughout our society for a new day in energy.

The rapidly developing photovoltaic technology is being whipsawed in these tides of debate while its solid technical base, now well established and growing, is ignored. There is no need to resort to "technological timidity in a country that is teeming with technological strength," as Daniel Greenberg put it.[11] The extensive development effort now under way, from basic research to engineering design—including rapid innovation in manufacturing techniques and full-scale tests of systems in the real world—virtually assures that affordable photovoltaic systems with conversion efficiencies approaching 10 percent will be on the market within a year or two. In many areas and for special uses, large-scale first-generation production units were in place by 1983, and photovoltaic systems will become competitive for residential use in some parts of the country within the next several years.

Photovoltaics is a rapidly evolving technology. New cell design concepts, new materials from which to make cells, steadily increased efficiencies, and simplified production techniques virtually assure the mass production of low-cost, long-lived cells.

Consider the following scenario:

- Costs will continue to decline, just as they have historically, with a pronounced drop occurring when automated factories begin high-volume production of thin film or ribbon arrays.
- Residential uses will grow at an exponential rate, and utilities and their customers will work out new relationships regarding the production, use, and pricing of electricity (an area in which there are already some interesting precedents). Simultaneously, the utilities will begin to experiment with their own photovoltaic collectors as these become relatively more attractive when measured against fossil and nuclear plants, which carry high capital, fuel, and operation and maintenance costs.

Figure 1-7 Photovoltaic desalination unit now in operation in Saudi Arabia. This 8-kilowatt unit pumps water from a shallow brackish well and desalts it by the reverse osmosis method. (Courtesy: Solar Design Associates, Lincoln, Mass.; photo by Mobil Oil Corp.)

- Homeowners will begin to experiment with energy storage systems to see them through the evening hours and the inevitable rainy days. (Here again, there are imaginative technological developments, including improved batteries and inertial storage in flywheels.)
- In addition to supplying electrical power for household uses, including air conditioning and space heating (normally by heat pump), photovoltaic systems will gain increasing popularity as the principal source of power for the family automobile. Five hundred square feet of photovoltaic array will take care of household power needs, except for air conditioning and heating. Double this array, and photovoltaics can recharge the battery for powering the family auto.
- Banks or financial institutions will be quite happy to lend money for a system guaranteed to last 20 or more years. Household insurance will routinely include wind, hail, or storm damage to the array.
- The customer will be able to service his own system, obtain service as needed, or buy an annual maintenance contract.

Within the next decade and a half—perhaps sooner—photovoltaic systems can be the preferred, most economic energy option for providing

Figure 1-8 The Solar Challenger, a photovoltaic-powered plane, had its first flight on November 20, 1980. Paul J. MacCready, president of Aerovironment, Inc. of Pasadena, California, constructed the plane to heighten people's awareness of the potential of photovoltaic cells. On July 7, 1981, it successfully flew across the English Channel, covering 180 miles from Paris to Manston, England, in 5 hours 22 minutes, reaching an altitude of 11,000 feet (3,353 meters). Average speed was 35 miles per hour. (Courtesy: Don Monroe)

electricity for homes, offices, schools, automobiles, and light industry, as well as for remote sites, miles from any central utility, throughout the United States and major parts of the world.[12]

How we proceed to develop photovoltaic energy is inevitably and intimately intertwined with the uses we choose to make of it.The future depends on our ability to reach imaginative solutions as the generation of power is decentralized and utilities ease away from their role as suppliers to assume more the role of brokers.

The stakes are very high, and we cannot wait until every aspect of the technology is completely ready for commercialization. Photovoltaic power and the changes it will bring are inevitable. We would be well advised to deal with these realities affirmatively and deal with them now.

We are face to face with a watershed decision on a new energy technology. Photovoltaic technology is no longer exotic, casting a distant rosy glow over the twenty-first century. We must begin to treat it as a practical here-and-now reality to be integrated fully into our decisions about energy supplies.

NOTES

1. "Energy Is Still the Problem," *New York Times,* 29 July 1979, p. 18-E.
2. Robert Stobaugh and Daniel Yergin, eds., *Energy Future: Report of the Energy Project at the Harvard Business School* (New York: Random House, 1979).
3. Terms not explained in the text are defined in the Glossary.
4. *An Assessment of Solar Energy as a National Energy Resource.* NSF/NASA Solar Energy Panel, December 1972. NSF RA-N-73-001 (Springfield, Va.: National Technical Information Service, 1973), p. 5. A year later, NSF sponsored a workshop conference at Cherry Hill, N.J. This three-day meeting was the seminal event that launched the federal effort to make photovoltaics economically useful for applications on the earth. See *Photovoltaic Conversion of Solar Energy for Terrestrial Applications,* Oct. 23–25, 1973, Vols. I and II, NSF-RA-N-74-013; Executive Report, NSF RA-N-74-073 (Springfield, Va.: National Technical Information Service, 1974).
5. U.S. Executive Office of the President, Council on Environmental Quality, *Solar Energy: Progress and Promise* (Washington, D.C.: Government Printing Office, April 1978), p. 6.
6. President's message on Solar Energy, 20 June 1979, Fact Sheet, pp. 1, 3. Also see *Domestic Policy Review of Solar Energy.* U.S. DOE TID 28834 (Springfield, Va.: National Technical Information Service, 1979), p. 50.
7. Stobaugh and Yergin, *Energy Future,* pp. 209–210.
8. ". . . If only a part of the $2 billion of federal funds slated for the Clinch River breeder reactor were directed instead to photovoltaic purchases, the $1,000-per-peak-kilowatt price could be achievable very soon—compared to a $5,000-per-peak-kilowatt estimated cost at Clinch River."
9. Nuclear fusion and breeder reactors, which are sometimes classified as renewable energy sources because they could for all practical purposes provide unlimited supplies, are not included in the popular concept.
10. "Solar Power Without Granola," *Washington Post,* 2 May 1979, editorial.
11. Daniel Greenberg, "Technological Timidity," *Washington Post,* 22 May 1979, p. A-15.
12. Photovoltaic systems can be designed to collect heat energy in addition to producing electricity. Such dual-purpose designs are sometimes called "total energy" systems.

FURTHER READING

Application of Solar Energy to Today's Energy Needs. Vols. I, II. U.S. Congress, Office of Technology Assessment. Washington, D.C.: Government Printing Office, 1975.

Commoner, Barry. "The Solar Transition." *New Yorker,* Part I: April 23, 1979; Part II: April 30, 1979.

Environmental Quality: The Ninth Annual Report of the Council on Environmental Quality. U.S. Executive Office of the President, Council on Environmental Quality, Washington, D.C.: Government Printing Office, 1978.

Hunt, V. Dan. *Solar Energy Dictionary.* New York: Industrial Press, 1982.

Lovins, Amory B. "Energy Strategy: The Road Not Taken?" *Foreign Affairs* 55, October 1976, pp. 65–96.

Schumacher, E. F. *Small Is Beautiful. Economics as if People Mattered.* New York: Harper & Row, 1973.

Stobaugh, Robert and Yergin, Daniel, eds. *Energy Future: Report of the Energy Project at the Harvard Business School.* New York: Random House, 1979.

Wilson, Carroll L., project director. *Energy: Global Prospects, 1985-2000—Report of the Workshop on Alternative Energy Strategies, Massachusetts Institute of Technology.* New York: McGraw-Hill, 1977.

CHAPTER 2
HOW SOLAR CELLS WORK

The photovoltaic phenomenon—the process by which light is converted silently and directly into electricity, without the elaborate machinery we usually associate with the generation of electricity—is elegant. It is certainly far less complicated than an atomic reactor, or for that matter, even a conventional coal- or oil-fired plant. A review of the science underlying photovoltaics will help to explain why photovoltaics is at once safe, clean, durable, reliable, energy-efficient, and increasingly socially and economically attractive.

LIGHT

Light drives the photovoltaic process and provides the energy that is translated into electric energy.

Light is composed of tiny bundles of energy, which act like individual bullets. These bundles have mass (weight), and they travel at extremely high, yet finite, measurable speed. Each bullet is called a photon, and the energy (momentum) a photon possesses is the product of its mass and speed, just as is the case for a bullet from a gun. A stream of photons—a beam of light—behaves in some ways like a string of waves, and for decades light was thought to be a special kind of wave motion in a mysterious hypothetical medium called the ether. The apparent length of these waves is shorter for photons of higher energy, and in fact, wavelength is one way of indicating energy content. More than a hundred years ago, a French scientist (Becquerel, in 1839) noticed that when light fell on one side of a very simple kind of battery cell it produced an electric current in a wire. For many years no one knew why. Years later, the mystery began to unravel when it was discovered that an atom consists of a minute nucleus surrounded by electrons. When a photon strikes an atom, it can interact with the electrons and be absorbed. The added energy can drive one of the atom's outer electrons off. (An

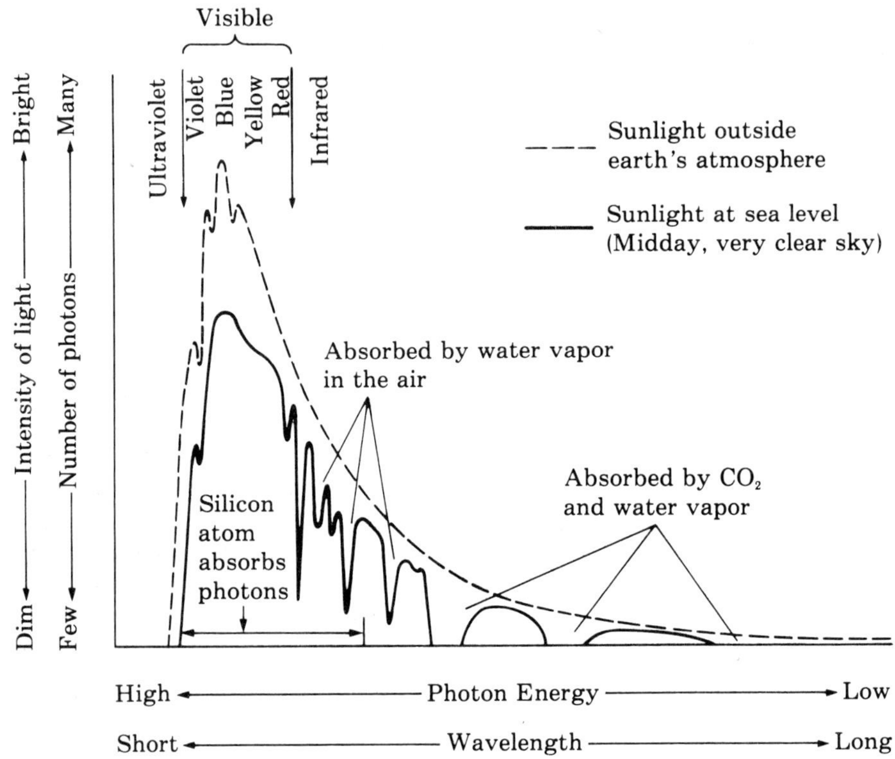

Figure 2-1 Solar spectrum. The area between the dotted line and solid line represents the portion of sunlight absorbed by the earth's atmosphere; some components of sunlight absorb strongly in certain regions. Our eyes are sensitive to the part of the spectrum where the photons are most numerous. A silicon atom absorbs photons in the region shown.

electron freed in this manner is known as a conduction electron because it is free to move and form an electric current.)

This interaction between light and electrons is at the heart of all photovoltaic devices. The energy being continually generated within the sun, whose surface is about 10,000°F (6,000°C) is liberated as a stream of photons of various energies leaving the surface. Indeed, because of this, the sun is slowly losing weight, or mass. Each photon reaching the earth has all of the energy it possessed when it left the sun's surface some 8½ minutes earlier. A simple picture-diagram of sunlight shows the distribution, or spectrum, of these energies and how they relate to the colors our eyes detect (Fig. 2-1).

In solar photovoltaics, two aspects of sunlight are important: how many photons reach a given spot on the earth and how much energy they have. As we suggested, sunshine is like a stream of bullets, all traveling at the same speed, some large and some small. Since energy is the product of speed times mass, the photons are distributed over a range of energies.

Above the earth's atmosphere, the energy in sunlight is about 1,358 Watts per square meter. At sea level, this energy is reduced by atmospheric attenuation to about 930 Watts per square meter. (A square meter contains 10.76 square feet, an area somewhat larger than 1 square yard.) Thus, on a clear day when the sun is directly overhead or nearly so, if all of the sun's energy falling on 1 square meter of the earth were converted into electricity, it would light ten 100-Watt bulbs. Unfortunately, it is not possible to convert even half of this solar energy into electricity. Nevertheless, the total energy being received is large enough that practical conversion devices are now becoming economically attractive.

About 30 percent of the energy reaching the earth's surface is immediately reflected into space as earthshine[1]. Another 47 percent is absorbed, at which instant the photons' energy is changed into heat.[2] (This is why sunlight feels warm on the back of your hand even on a cold day.) Indeed, photons must be absorbed for heat to be released. In terms of heat energy, 1 square meter of full sunlight at high noon yields about 750 calories, or about 3,000 Btu's, every hour (enough, for example, to increase the temperature of 10 gallons of water by 36°F).

SINGLE-CRYSTAL SILICON

The material most commonly used to make photovoltaic converters is silicon. Silicon is a good photovoltaic material because most of the incoming photons will displace electrons. All of the photons with energy greater than 1.08 electron Volts (corresponding to a wavelength of 11,500 Angstroms, in the infrared) will be absorbed and activate an electron. Therefore, it appears that nearly all of the energy in sunlight could be converted to electricity. Unfortunately, this turns out not to be the case because one photon can trigger the release of only one electron. Any photon that has more energy than the minimum required for releasing one electron has its remaining energy converted to heat in the silicon.

The net effect is that silicon solar cells can at best convert about 25 percent of the total energy in sunlight into electricity. This is a theoretical limit imposed by nature's building blocks and our ability to build practical conversion devices. In some cases, it may be profitable to build systems that capture this excess heat energy for practical applications where both electricity and heat are needed. Such combined systems are called photovoltaic-thermal or "total energy" systems.

There are other inefficiencies in practical silicon-based systems that further limit the amount of electricity that can be produced. First, some sunlight never gets into the cell because it is reflected from the surface. This loss can be partially overcome by using antireflective coatings. Second, the wire net or grid used in the construction of the cell blocks out some incoming sunlight. Third, depending on the purity of the silicon and the refinement of the crystal structure, there are internal losses of energy. And, finally, in a photovoltaic array all of the area devoted to the collection of sunlight is not fully covered by solar cells. While this is minimized by making the cells rectangular or hexagonal rather than circular, as most early cells were, some space is still not used. For silicon systems, these inefficiencies are given in Table 2-1.

Table 2-1 Inefficiencies of Silicon Photovoltaic Modules

Factors Contributing to Energy Loss	*Percentage of Sunlight Not Converted into Electricity*
Overly energetic photons	32
Under-energetic photons	24
Other internal cell losses	21
Front surface reflection	3
Shading by contact bar and fingers	3
Cell packing density	2
Total	85

Good silicon photovoltaic systems for sale today have an overall conversion efficiency of about 10–15 percent.

Let's look at the basic properties of matter that the photovoltaic phenomenon capitalizes on. The simplest atom is hydrogen, which has 1 proton for a nucleus and 1 electron in orbit around it. The largest atom found in nature in any appreciable quantity is uranium, which has 92 electrons in orbit around a nucleus of protons and neutrons. Beginning with hydrogen and moving towards larger atoms of heavier elements, the number of electrons builds up around the nucleus in successive layers or shells. The outer shell in the smaller atoms is full if it contains 8 electrons; the addition of another electron starts a new shell (Fig. 2-2).

An electron is a minute bit of matter compared to the atoms of which it forms a part. It has a mass about $1/1{,}840$ that of a single proton or a single

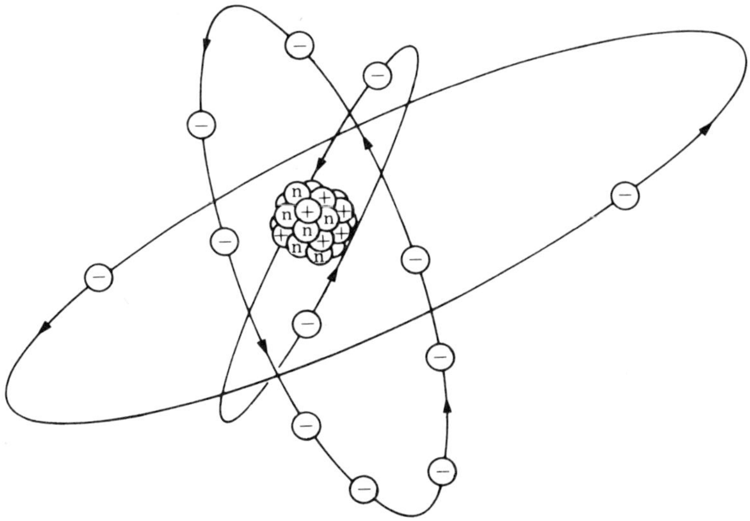

Figure 2-2 Magnesium atom. One of the smaller atoms is magnesium, which is metallic. The nucleus is a tiny cluster of 12 protons ⊕ and 12 neutrons ⓝ. The elliptical orbits around the nucleus (in three dimensions) are the successive layers, or shells, of electrons ⊖. The innermost shell of all atoms is unique in that it is filled by only 2 electrons.

Each protron has one positive charge, each electron one negative charge. Every atom has the same number of electrons as there are protons, so that the atom as a whole is electrically neutral. Neutrons carry no charge.

In metals, such as magnesium, the electrons in the outer shell are not strongly attached, moving easily from one atom to another. For this reason, metals are good conductors of electricity, which is merely a flow of these outer shell electrons.

The silicon atom has 14 protons and 14 neutrons in the nucleus and 4 electrons in the outmost shell.

neutron, both of which compose the nucleus of an atom. If the outer shell of an atom contains only one or two electrons, these are loosely held and move easily from one atom to another. Elements composed of such atoms are metallic and readily conduct electricity—which is itself nothing more than a flow of these outer shell electrons. (Indeed, a wire conductor can be likened to a pipe full of water: If additional electrons are pushed in one end, the electrons already in the wire tend to be forced out the other end, or out at any point that they can find an escape route. A generator is merely a device that pushes electrons through a wire.)

On the other hand, if the outer shell of an atom contains six or seven electrons, the material is non-metallic and a poor conductor of electricity because the electrons are tightly bound.

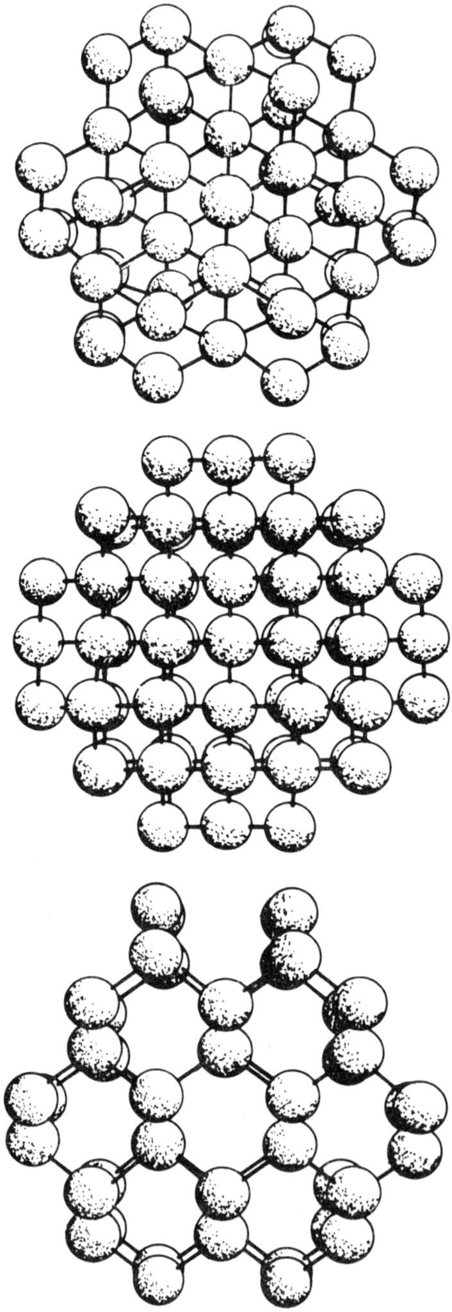

Figure 2-3 Model of a silicon crystal. Each atom is firmly positioned by electron bonds to four surrounding atoms. This model is shown from several angles.

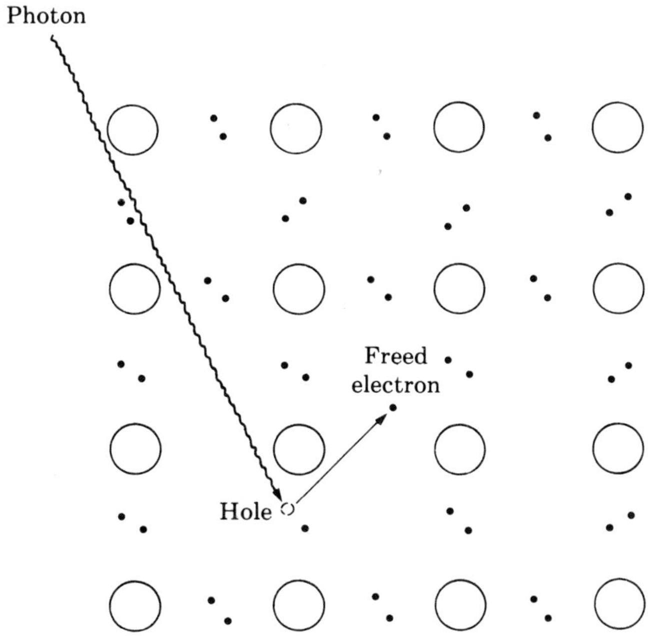

Figure 2-4 Silicon crystal. When molten silicon cools very slowly, the atoms (large circles) align themselves in a perfect tetrahedral pattern. Each atom has four outer shell electrons (black dots), which are shared with a neighboring atom to provide a stable structure (shown here in two dimensions). A photon (particle of light) is shown freeing an electron.

Between these extremes are elements whose atoms have three, four, or five electrons in the outer shell. These outer electrons can be freed, but only if given some additional energy, or push, from outside. Because they conduct electricity slightly, such elements are called semiconductors. An atom of silicon, a principal constituent of sand, has four outer shell, or valence, electrons.

A silicon atom readily absorbs a photon, and the added energy will "excite" or activate one of the outer electrons and thus free it. This is best accomplished when silicon atoms are lined up in precise rows or positions. This is called the crystalline state. In any crystal, the atoms or molecules are arranged in perfect geometrical formation (Fig. 2-3). (The opposite is the amorphous state, in which atoms or molecules are jumbled together in no regular pattern.)

When a photon strikes this silicon crystal, it penetrates until absorbed by an atom (Fig. 2-4). Almost immediately the photon's energy is transferred to one of the outer electrons, which then breaks free from the atom and bounces away through the crystal lattice leaving a vacant position, or hole, where it was. When a beam of light falls on a silicon crystal, millions of electrons are freed in this manner, and an equal number of holes are left in the lattice. Any electron that happens upon any hole may refill it, giving off a minute amount of heat in the process. When the light is shut off, all the displaced electrons are quickly reabsorbed in the holes. The crystal returns to its original state, becoming slightly warmer. Because the atoms do not change position, the crystal itself remains completely unaltered in size, weight, form, and appearance.

A trick has been discovered that upsets this orderly state of affairs for a useful purpose. By introducing a few foreign atoms into a silicon crystal, the perfect arrangement of outer shell electrons can be upset or disordered. A crystal can be created that has either too many electrons or not quite enough to satisfy its internal structural balance. Two elements are typically used for this purpose: boron, which has three outer shell electrons, or phosphorus, which has five. Only a minute amount of these impurities is needed, perhaps one part in a million.

Whenever a boron atom is added, it leaves a vacancy of one electron in the lattice. Thus, a silicon crystal "doped" with boron needs electrons to make the lattice complete, and it tends to absorb electrons much as a sponge soaks up water. On the other hand, silicon doped with phosphorus atoms, each of which has five outer shell electrons, will have an excess of electrons in the lattice.

HOW A PHOTOVOLTAIC CELL WORKS

A solar cell is made by placing a thin layer of phosphorus-doped silicon in intimate contact with a layer of boron-doped silicon. When light falls on the cell, photons are absorbed and electrons are set free. The excess electrons accumulate in the phosphorus-doped silicon, which is called n-silicon because electrons have negative charge. If one end of a wire is attached to this top layer and the other end connected to the layer beneath, electrons will leave the upper layer, flow through the wire, and be absorbed by the boron-doped silicon, which is called p-silicon, meaning positive (Fig. 2-5).

The electrons flowing through the external wire circuit can be used to light a light or turn a motor, just as electricity from any other source. Thus, the solar photovoltaic cell, using light only, generates electricity.

Why do the free electrons prefer to leave via the wire rather than be absorbed in the holes they have left behind in the crystal? When the cell is made, a peculiar phenomenon occurs at the plane of contact between the two silicon layers. Immediately, some of the excess electrons in the n-layer diffuse a short distance across the interface into the p-layer just beneath.

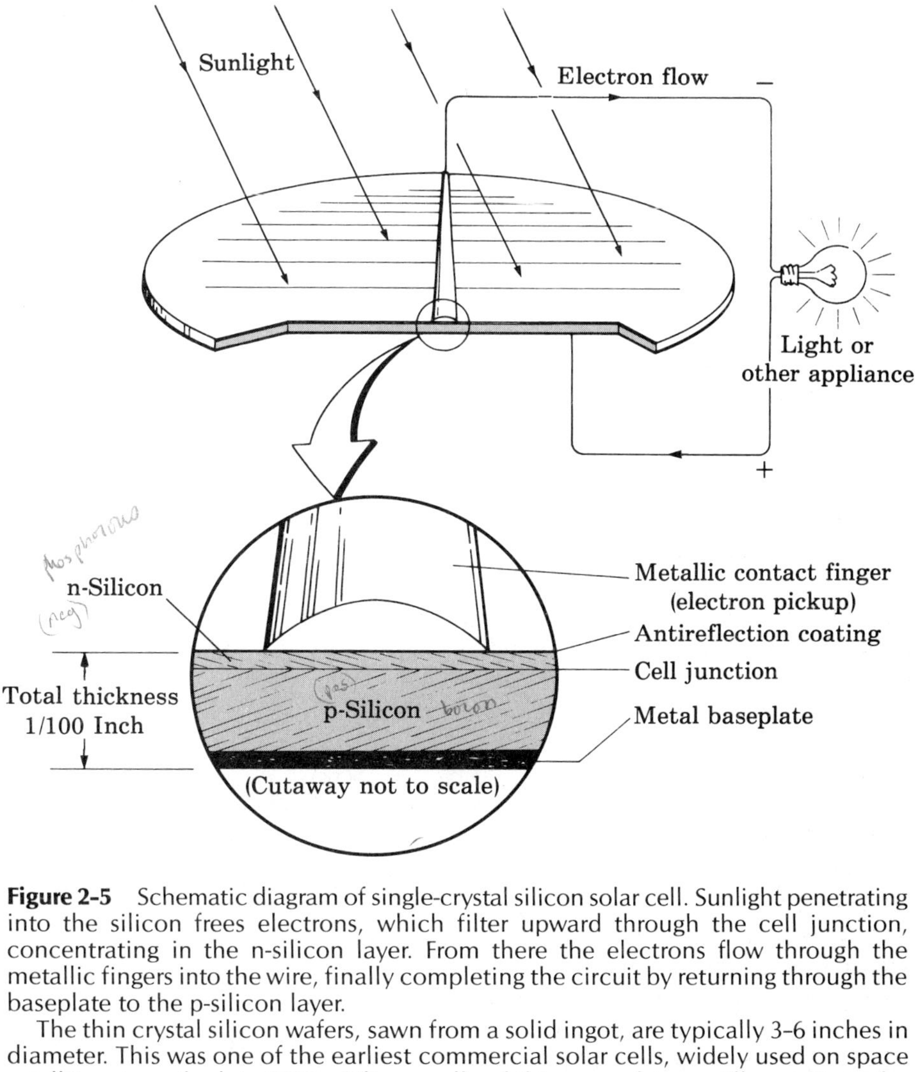

Figure 2-5 Schematic diagram of single-crystal silicon solar cell. Sunlight penetrating into the silicon frees electrons, which filter upward through the cell junction, concentrating in the n-silicon layer. From there the electrons flow through the metallic fingers into the wire, finally completing the circuit by returning through the baseplate to the p-silicon layer.

The thin crystal silicon wafers, sawn from a solid ingot, are typically 3–6 inches in diameter. This was one of the earliest commercial solar cells, widely used on space satellites since the late 1950s. Silicon cells of this general type still constitute the mainstay of the commercial solar photovoltaic business, although now rectangular in shape in order to provide more compact arrays.

They are attracted by the holes there, which they "want" to fill. This leaves the phosphorus atoms in the n-layer without enough electrons to balance fully the positive charge in their nuclei, while too many electrons are located in the p-layer around the boron atoms. A very thin layer of static electrical

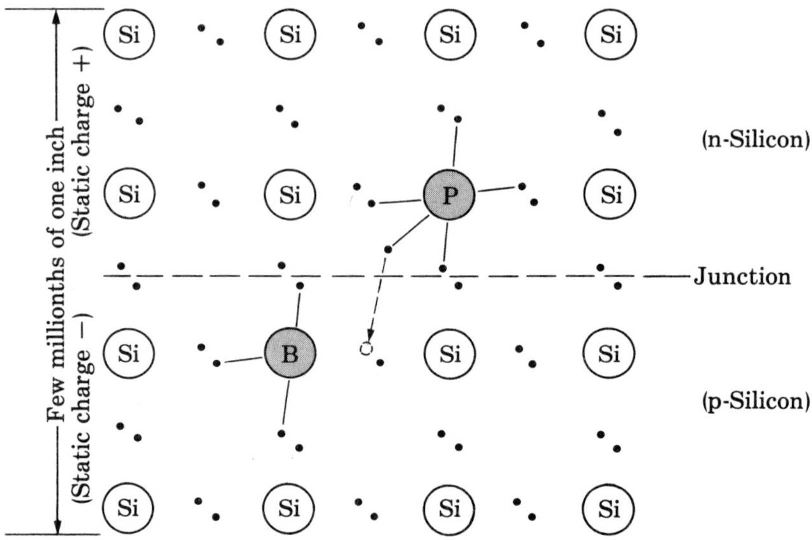

Figure 2-6 How the cell barrier is established. Open circles are silicon atoms; black dots are electrons. The neutral phosphorus atom, P, originally has five electrons. The neutral boron atom, B, has only three, leaving a hole in the lattice structure. The extra electron from the phosphorus atom immediately moves to fill it. The phosphorus atom is then left with a net positive charge, and the boron atom picks up the negative charge of the added electron. No light (photon) is involved in this step, which occurs spontaneously.

This exchange occurs only in a microscopically thin region where the two oppositely doped layers are in contact. A region of static electrical charge can also be established between two dissimilar semiconductor materials or between a semiconductor and a metal.

charge is thus formed along the zone of contact, or junction, between the two layers of silicon (Fig. 2-6).

Such static electricity is a commonplace phenomenon. If you rub your hand hard across your hair, some electrons from the hair atoms are rubbed off and stick to the palm of your hand. If you then hold your hand close to your hair, hair will rise toward it because the hair, now positively charged, is attracted by the negative charge on your hand. A sweater taken off suddenly will often pick up enough electric charge so that, in a dark room, you can see sparks jump from it when it is laid down. Clothes tumbling in a dryer develop strong static electricity as electrons are rubbed off one fabric by another. Such effects are more noticeable in dry weather when the air will not conduct the electrons back to their point of origin as readily.

Because electrons are charged particles, it is difficult for them to pass through the zone of static electric charge in a solar cell. For this reason, the charged region is called the cell barrier. When a cell is fabricated, the barrier establishes itself instantly. It lasts for the life of the cell. It never wears out. The cell barrier is extraordinarily thin, calculated to be about one-millionth of an inch thick. Every solar cell must have such a barrier, or zone of static charge. No solar cell would work without it.[3]

Because the barrier resists the passage of electrons through it, only electrons with high energy (great speed) can penetrate it. As a result the barrier acts as a filter that lets high-energy electrons through and stops low-energy electrons.

An analogy can help explain how this happens. A solar cell can be likened to two tables with marbles on them (Fig. 2-7). The two tables are connected by a small elevation or ridge over which marbles can move if they have enough speed. Table A on the left is vibrated gently (in the horizontal plane) so that the marbles are jostled about, and occasionally a marble will roll over the ridge to Table B. On the other hand, Table B is vibrated forcefully so that frequently marbles will have enough energy to get over the ridge in the other direction to Table A. Eventually, as marbles accumulate on Table A and the number remaining on Table B diminishes, the number of marbles crossing the ridge will be the same in each direction, and a condition of equilibrium will be established. However, if the two tables are connected by an external channel, which we will assume is flexible, the press of marbles on A will cause some of them to be pushed through this channel back to Table B.

In a solar cell, the p–n junction is like the ridge between the tables. The free electrons act like the marbles. The p-silicon lattice, corresponding to Table B, has an excess of holes due to the boron dopant, and these holes tend to absorb electrons. As a result, there are not many free electrons in the p-silicon, but those that are free move at high speeds (have high energy).

The opposite is true in the n-silicon, which corresponds to Table A. Here there are not enough holes, and there are too many electrons, so that the average electron has less energy (is moving slower) than those in the p-silicon. Electrons in both layers are moving at random, some driving by chance into the barrier. The higher-energy ones from the p-silicon penetrate the barrier into the n-silicon, while low-energy electrons in the n-layer are prevented from returning. Thus voltage, or electron pressure, is created between the two layers. If a wire is connected externally from the n-layer to the p-layer, the excess electrons will readily flow through it.

As electrons migrate preferentially into the n-layer, the holes in the lattice, of course, can be visualized as progressively "moving" in the opposite direction. Electrons coming back through the external wire eventually end up being absorbed in holes at the base of the p-layer, which in effect were

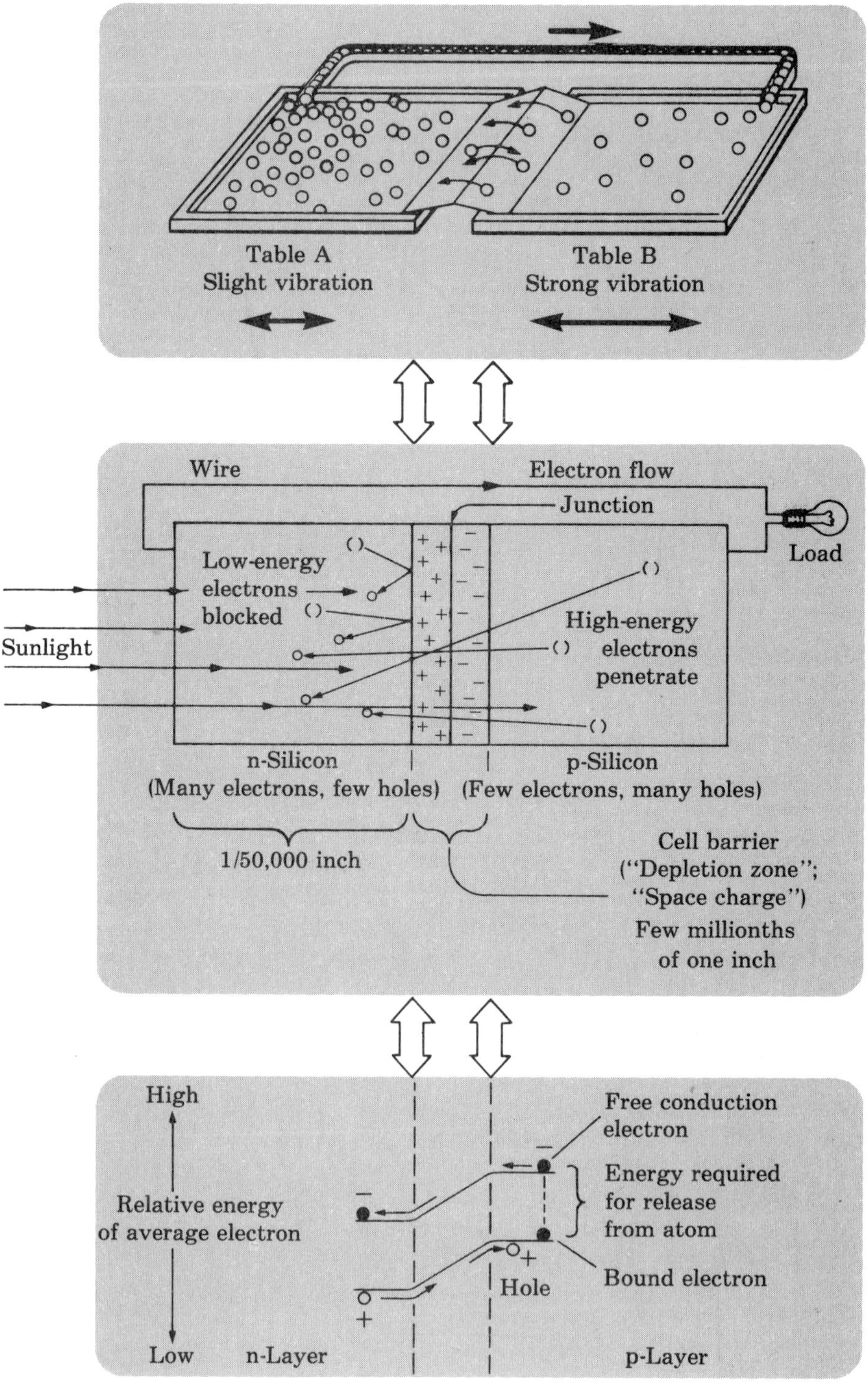
Table A
Slight vibration
Table B
Strong vibration
Wire
Electron flow
Junction
Load
Low-energy electrons blocked
High-energy electrons penetrate
Sunlight
n-Silicon
(Many electrons, few holes)
p-Silicon
(Few electrons, many holes)
1/50,000 inch
Cell barrier
("Depletion zone";
"Space charge")
Few millionths of one inch
High
Relative energy of average electron
Low
Free conduction electron
Energy required for release from atom
Bound electron
Hole
n-Layer
p-Layer

originally produced when the electrons broke free. Electrons move so rapidly that if the source of light is shut off, the free electrons are almost instantly reabsorbed in holes and the process comes very quickly to a halt.

The incoming photons keep the process going by continually creating new electron-hole pairs. Indeed, it does not matter which side of the junction the new pairs are created on. It is important only that they be created in the vicinity of the junction. If a photon is absorbed in the n-layer, the electron that is freed simply adds to the number already present on that side, which is desirable; the hole migrates eventually through the junction as incoming electrons successively fill it. If a photon is absorbed in the p-layer, the process is as described above.

The n-layer is usually made of such thickness that the junction will be at optimum depth for absorbing photons. For silicon, this is about 0.5 micrometer (1/50,000 of an inch).

The entire thickness of the cell need be no greater than 250 micrometers (0.25 millimeter), about 1/100 of an inch. This is only three times the thickness of a human hair. It is theoretically possible to make solar cells that are extraordinarily thin, and thus much cheaper, especially in mass production.

To portray what is happening in a solar cell, the physicist resorts to a diagram as shown in Figure 2-7. Here the cell is imagined to be in the position of the two tables of marbles, with the n-silicon on the left. Like the ridge between the tables, the junction or barrier is in the middle. The physicist uses the vertical dimension in the diagram to portray the relative energy of both the free electrons and the bound electrons in each region of the cell. The free or "conduction" electrons, represented by the upper line, move from the region of higher average energy in the p-silicon on the right, through

Figure 2-7 How a solar cell works. In this diagram, the solar cell is pictured (center) lying on its side with the sunlight entering from the left. At the center is the cell barrier—also known as the depletion zone, or space charge. This is a thin region of static electric charge, which resists the passage of electrons. High-energy electrons on the right are able to penetrate this region; low-energy electrons on the left are repelled. The process can be pictured as basically mechanistic, as depicted by the table of marbles (top). Just as the marbles accumulate preferentially on Table A, so the electrons accumulate in the n-silicon.

In the cell (center) the free electrons bounce about at random, colliding frequently and changing directions with every collision. Thus the long, straight paths, shown here to illustrate the principle, are an oversimplification.

The physicist uses an energy diagram (bottom) to portray the movement of electrons from a high-energy state (right) through the barrier to a lower-energy state (left), while an equal number of lattice vacancies migrates in the opposite direction.

the barrier, to the region of lower average energy in the n-silicon on the left. Simultaneously, the holes or vacancies migrate in the opposite direction. Indeed, the explanation is frequently given that the electrons in the p-layer near the junction are "sucked down" through the barrier to the zone of lower energy.[4]

Particle physics sees the matter a bit differently. The process can be envisioned as basically mechanistic. The photovoltaic phenomenon is remarkable in that here electrons—the subatomic particles that constitute an electric current when they are flowing in unison in one direction—are behaving more like the bits of matter that they are. Within the crystal, they are diffusing through the cell barrier, much as molecules of gas might diffuse through a porous barrier. But because these are electrons in motion, their motion does in fact conform with electrical theory, and the process is often explained in that context. Thus the holes, which are regions of positive electrical charge, are pictured as "moving" in the opposite direction from the negatively charged electrons, in order to make things come out even from the standpoint of electrical theory.

It is similar to saying that when automobiles move from the city to the suburbs in the evening, an equal number of empty parking spaces "flows" into the city. If each driver had several errands to perform and therefore made a number of stops on the way home, occupying a succession of parking spaces, then the picture of empty spaces flowing toward the city as cars flowed homeward takes on more realism.

The electronics engineer uses the flow of electrons and of holes to describe what is happening in a photovoltaic cell because the lattice vacancies are regions of positive charge, which in his system of accounting, must exactly equal the number of negative charges carried by the moving electrons. But in terms of movement of particles of matter, only the electrons are involved.

The sole function of the photons is to keep replenishing the supply of free electrons by energizing electrons from their trapped positions in orbit about the nuclei into a free state, called the conduction band. The cell does the rest of the job. The photons, in effect, provide a pumping action, and the energy they supply is converted into the energy of electrons in motion, and thus into an electric current that lights a light, drives a motor, or produces heat.

The amount of current (amperage) produced by a photovoltaic cell is proportional to the amount of light falling on the cell (the number of photons entering it). For this reason, current increases with the area of the cell as well as with the intensity of the light. The voltage, on the other hand, depends on the materials used. All silicon cells produce about ½ Volt regardless of cell area. A single-cell flashlight battery is rated at 1½ Volts, which can be provided by three silicon photovoltaic cells connected in series so that their voltages are additive.

Solar cells are unique energy producers in that no materials are consumed or given off. For this reason solar photovoltaic receivers can be completely sealed units that, in principle, are as durable as the materials from which they are made. In this important respect, photovoltaic devices are very different from batteries. The cell materials in batteries must undergo change in chemical composition, during which the atoms, in the form of charged ions, move about. For example, lead may dissolve in acid as electricity is withdrawn and be plated out again as the cell is recharged. Because the material itself changes form, it never is replaced exactly as it was, and even the best batteries in time simply wear out.

But in photovoltaic cells, the atoms themselves do not change position; only the electrons between them move. The material does not change shape or form any more than, say, a copper wire that conducts electricity for years without undergoing any change. It is possible to make solar cells that operate for 20 or 30 years with very little loss of effectiveness. Energy from light is simply being converted directly into electrical energy for external use.

POLYCRYSTALLINE SILICON

Other forms of silicon that are much cheaper to produce than large single crystals will also work in photovoltaic applications. When molten silicon cools under normal circumstances, the resulting ingot consists of a myriad of microscopic crystals. If the cooling is done very slowly, the crystallites will be larger in size. The product, called polycrystalline silicon (polysilicon or poly) is granular in nature, and the size of the crystallites depends on the conditions under which the silicon cooled. Polysilicon can be formed as an ingot, or solid block, or in the shape of a ribbon drawn up from a molten bath or by boiling the silicon off and allowing the vapor to deposit and cool on some flat surface. These methods are much less expensive than growing single crystal silicon.

Silicon cells made from polycrystalline silicon are somewhat less efficient in converting sunlight into electricity. Minute electrical shorts occur along the grain boundaries between the crystallites. The larger the crystallites, therefore, the fewer the boundaries and the more the resulting cell behaves as if it were made from a single crystal (Fig. 2-8).

In general, it is desirable to have the individual crystals much longer than the thickness of the cell in order to reduce losses from electrical shorts along the grain boundary. Grains ranging in size from about 0.10 millimeter to several millimeters in length can be made, and they work quite satisfactorily.

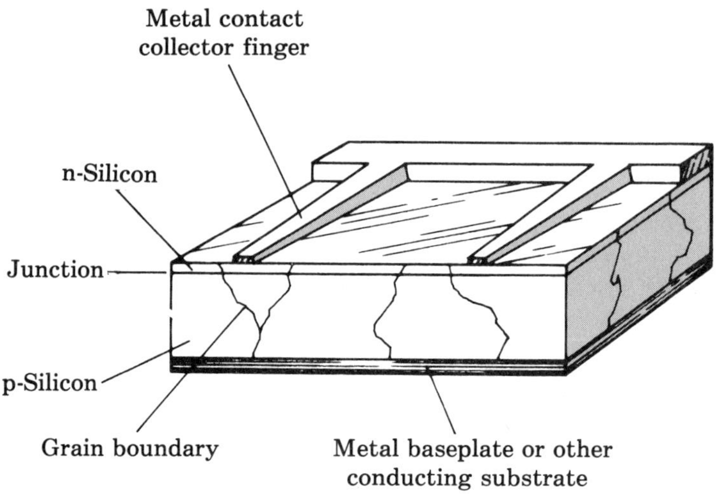

Figure 2-8 Cross-sectional diagram of silicon cell made of polycrystalline material. Free electrons find conducting pathways along the sides of the crystallites, enabling them to recombine with holes before passing through the cell junction. In addition, electrons in the n-layer are inhibited from moving across crystallite boundaries to reach the metallic grid fingers on top. Thus the cell is partially shorted out internally. This problem can be reduced in two ways: by providing larger crystallites, which reduces the number of grain boundaries; and by treating the silicon with hydrogen gas, which is absorbed within the cracks and, for reasons not yet fully understood, reduces the unwanted conductivity of electrons.

One way of reducing electrical seepage along the boundary is to treat the tiny crystals with some material that, in effect, insulates the edges of the grains. Hydrogen is being used for this purpose. Cell efficiencies as high as 10 percent are being achieved for polysilicon, depending on the design of the cell.

Polycrystalline silicon grown in ingot form still has to be sawed into thin wafers, an expensive step required in making cells from single-crystal material. But polysilicon can now be formed in a thin sheet in the first place, by vapor deposition or by simply dipping a substrate, such as a piece of graphite, in a bath of molten silicon, eliminating the sawing step. Such procedures, which are being developed vigorously, are more adaptable to mass production techniques and should make it easier to produce cells that are larger in area. The goal is to reduce cell cost without sacrificing conversion efficiency, reliability, or the useful lifetime of the product.

AMORPHOUS SILICON

Even amorphous silicon, the noncrystalline material (also called alpha-silicon) can be used to make solar cells. Amorphous (without form) silicon is pure silicon that has no crystal properties. The silicon atoms in the solid are randomly distributed. Glass, for example, is amorphous, not crystalline, material. Very thin films of amorphous silicon can convert sunlight into electricity. Since a very small amount of material is used and the crystallization step is avoided, amorphous silicon cells are very inexpensive.

The light energy conversion efficiency of amorphous silicon, however, is much lower: typically, 9–11 percent, compared with 16–18 percent for single-crystal cells. When solid state theory was developed, it centered on the structure and behavior of individual atoms and the forces that bind individual atoms together. It was easy to extend this knowledge to crystals because crystals are perfect geometrical arrangements of atoms. Because there was long-range order, what applied to the smallest unit applied to the whole crystal. The behavior of amorphous materials is much more complex, and basic knowledge of the structure and performance of amorphous material is limited. Better understanding of how matter performs in such conglomerations is now unfolding, and the long-term future of amorphous materials for power production is promising.[5]

Amorphous silicon cells are appearing in great numbers in Japanese products, powering small devices such as watches, hand calculators, and toys that need very small amounts of energy. Japanese firms are leading producers of amorphous silicon. If techniques can be devised for raising their efficiency a few percentage points and increasing their stability, amorphous silicon cells may in time become competitive for generating significant amounts of power in applications where the total area required for the collector surface is not critical.

OTHER KINDS OF SOLAR CELLS

Solar cells can be made from a number of materials other than silicon and formed in a variety of designs. Cells are classified both by material and by the type of juction, or barrier, they employ.

There are four basic kinds of junctions. In the silicon cell, the junction is the very thin region at the boundary between the phosphorus-doped and the boron-doped portions. The basic material is all silicon, and such a junction within a single material is called a *homojunction*. Solar cells can, however, be made of dissimilar materials. One type of solar cell has a layer of copper sulfide deposited on a layer of cadmium sulfide, and the junction forms along the contact between the two substances. Where such dissimilar materials are used, the contact zone is called a *heterojunction*. A junction can also be established between a semiconductor material and a piece of metal, and this type is called a Schottky barrier, for its discoverer. Finally, in a metal-insulator-semiconductor or MIS junction, a very thin layer (less than 0.003 micrometers) of some material such as titanium oxide is sandwiched between the metal and the semiconductor. A modification of the MIS cell is the semiconductor-insulator-semiconductor or SIS cell. Each of these basic types of junctions offers various advantages and has associated limitations.

Some of the more promising semiconductor materials and their theoretical efficiencies for converting sunlight into electricity are given in Table 2-2.[6]

Table 2-2 Efficiencies of Semiconductor Materials

Material	*Maximum Conversion Efficiency, Percent*
Germanium	13
Cadmium sulfide	18
Silicon	25
Cadmium telluride	25
Indium phosphide	26
Gallium arsenide	27
Aluminum-antimonide	27

Acknowledging there are many reasons "that rule out any hope of reaching more than about four-fifths of these efficiencies," Bruce Chalmers of Harvard University pointedly adds, "Lest it be thought that an efficiency of less than 20 percent is unworthy of our technical sophistication, it should be remembered that the epitome of advanced industrial technology, the automobile, performs its energy conversion at an efficiency of less than 20 percent, and it performs its people-moving function at a far lower figure."[7] In generating electricity from fossil fuels, two-thirds of the energy is lost in the conversion process.

Figure 2-9 Distinctively different from silicon cells are cadmium sulfide-copper sulfide cells. Though less efficient than silicon cells, they are extremely thin and readily adaptable to mass production. Two types are shown. The Front Wall cell is made by depositing the required materials in successive layers on a thin sheet of copper. In the Back Wall type, materials are deposited on a pane of glass. Sunlight enters through the glass, which provides weather protection. The most expensive part of this cell is simply the ⅛-inch-thick glass substrate.

Figures in parentheses are thicknesses in micrometers (one micrometer is one-millionth of a meter, or about 1/25,000 of an inch).

Front Wall Cell

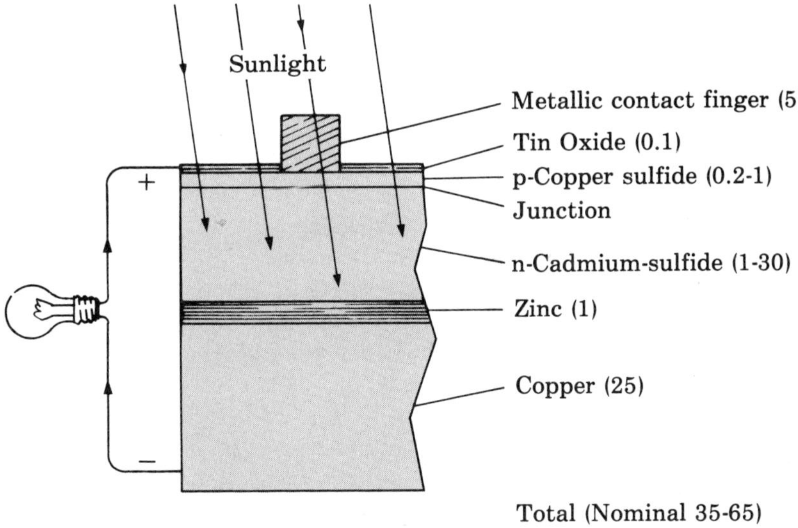

Back Wall Cell

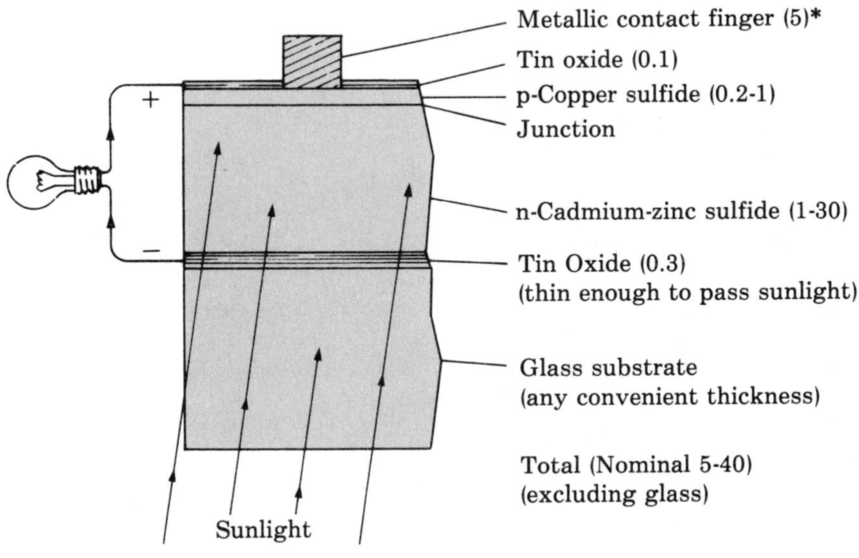

Extensive theoretical and applied research is in progress on these and a number of other materials. Depending upon the substances used and the type of junction employed, literally dozens of different kinds of solar cells are possible. The choice of cell material and the design of the cell influence not only the cell efficiency but also the fabrication method, which in turn affects costs. The worth of a cell, which is usually expressed in dollars per Watt of generating capacity, incorporates both factors.

Other than silicon, the materials most often used in solar cells today are cadmium sulfide and gallium arsenide. Cadmium sulfide cells are relatively easy to manufacture, they are inexpensive, and they have achieved conversion efficiencies of about 8 percent. Depending on how it is made, the cell proper may range in thickness from as much as 65 micrometers to as little as 5 micrometers. A metal substrate such as copper can be used, but more recently ⅛-inch-thick (0.3 centimeter) glass panes are being used (Fig. 2-9). In production, the most expensive part of the completed product is the glass backing or substrate. Copper sulfide cells have not shown adequate stability, however, and most manufacturers have ceased making them.

Gallium arsenide is important for two reasons. First, it is inherently more efficient because it captures a wider part of the solar spectrum than cells of other materials. Second, at high temperatures, it does not lose efficiency as rapidly as silicon. Gallium arsenide is one of the best materials known for use in concentrator cells, which focus the relatively diffuse sunlight on a small area and produce high temperatures. Moreover, this material may be produced either in a single-crystal form or as a polycrystalline thin film. Efficiencies approaching 25 percent have been demonstrated with single-crystal gallium arsenide cells, while 15 percent conversion has been attained with thin-film gallium arsenide devices only 1–2 micrometers thick.

Closely akin to amorphous silicon are the so-called chalcogenide glasses. These are semiconductor substances that contain large amounts of sulfur, selenium, or tellurium. Solar cells can be made of these materials today, but their efficiencies are low. The future, however, may hold another story.

As more is learned of the substances that make good solar cells and the arrangement of the atoms in such materials, chemists may be able to synthesize new compounds better tailored to specific requirements of photovoltaic systems. Vigorous research is under way on conductive polymers, plastics with interesting electrical properties.[8] This field is still in its infancy.

In summary, although photovoltaic theory is complex, the phenomena involved are commonplace daily experiences—as familiar as the warmth of sunshine on the back of your hand, the static electric charge in your hair, and the handy electric plug on the wall. There are no strange, new, unseen forces that society must try to comprehend.

The sheer elegance of the process is striking. There are no moving parts and no noise. In operation, nothing is consumed other than sunlight, and the output is high-quality energy in the form of versatile electricity. Because there are no material by-products, there is no pollution.

Photovoltaics is distinguished from the other solar technologies by the opportunities it holds for further discoveries of a fundamental nature. By comparison, direct collection of solar heat and the indirect methods of tapping solar energy, such as wind, ocean thermal systems, and the use of biomass, are limited for the most part to engineering improvements and simplification of production procedures. There is less chance for significant scientific breakthroughs.

The photovoltaic process is now well understood. The research is probing a variety of collateral issues—the effects of grain sizes, impurities, and temperature; grain boundary effects; properties of dozens of new materials; and many cell designs.

NOTES

1. A. L. Hammond, W. D. Metz, and T. H. Maugh II, *Energy and the Future* (Washington, D. C.: American Association for Advancement of Science, 1973), p. 49.
2. Of the remaining energy, close to 23 percent goes to evaporation of water, precipitation, etc. A relatively small amount (less than 1 percent) goes to wind, waves, and photosynthesis.
3. This zone of static electricity is sometimes called the "space charge." It is also referred to as the "depletion zone" because it is devoid of both free electrons in the n-layer and of holes in the p-layer since the electrons have been absorbed by the holes.
4. Another explanation sometimes given is that the cell barrier "separates" new electron-hole pairs as they are created by photon activity. This is the net result, but the picture is rather strained.
5. For an excellent discussion, see David Adler, "Amorphous Semiconductor Devices," *Scientific American* (May 1977), pp. 36–48.
6. An excellent overview of the leading photovoltaic materials from the standpoint of cell efficiency is contained in "Photovoltaic Materials," an article by Evelio A. Perez-Albuerne and Yuan-Sheng Tyan, *Science* 208 (23 May 1980), pp. 902–907.
7. Bruce Chalmers, "The Photovoltaic Generation of Electricity," *Scientific American* (October 1976), p. 39.
8. J. Mort, "Conductive Polymers," *Science* 208 (23 May 1980), pp. 819–825.

FURTHER READING

Adler, David. "Amorphous Semiconductor Devices." *Scientific American,* May 1977, pp. 36–48.

Chalmers, Bruce, "The Photovoltaic Generation of Electricity." *Scientific American,* October 1976, pp. 34–43.

Daniels, F. *Direct Use of the Sun's Energy.* New York: Ballantine, 1964.

Fan, John C. C. "Solar Cells: Plugging into the Sun." *Technology Review* 80, August-September 1978, pp. 2-19.

Five Year Research Plan, 1984-1988. Washington, D.C.: U.S. Department of Energy (Photovoltaic Energy Technology Division), May 1983.

Hovel, H. J. "Novel Materials and Devices for Sunlight Concentrating Systems." *IBM Journal of Research and Development* 22, March 1978, pp. 112-121.

Iversen, W. R. "Amorphous Silicon Wins Place in the Sun." *Electronics,* April 5, 1984, pp. 94-96.

Landsberg, P. T. "An Introduction to the Theory of Photovoltaic Cells." *Solid-State Electronics* 18, 1975, pp. 1043-1052.

McDaniels, D. K. *The Sun: Our Future Energy Source.* New York: John Wiley and Sons, 1979.

Neville, Richard C. *Solar Energy Conversion: The Solar Cell.* New York: Elsevier Scientific Publishing Co., 1978.

Pulfrey, David L. *Photovoltaic Power Generation.* New York: Van Nostrand Reinhold, 1978.

"SERI Photovoltaic Advanced Research and Development: FY 1983 Accomplishments." Golden, Colo.: Solar Energy Research Institute, March 1984.

"SERI Photovoltaic Advanced Research and Development—An Overview." Golden, Colo.: Solar Energy Research Institute, February 1984.

Sittig, Marshall. *Solar Cells for Photovoltaic Generation of Electricity—Materials, Devices and Applications.* Park Ridge, N.J.: Noyes Data Corp., 1979.

CHAPTER 3
HOW SOLAR CELLS ARE MADE

If price were no object, photovoltaics would already be widely used as a clean, safe, reliable, silent source of electric power. Price is indeed an object, however, as anyone who has studied his or her heating or electric bill and searched for alternative energy sources will agree. The price of photovoltaics is interwined with the technology for producing photovoltaic cells. Genuinely economical photovoltaic systems depend on the successful development of new production methods that capitalize on economies of scale through automation and the development of more energy-efficient collectors. Soon to be cost-competitive, photovoltaics can be expected to reorder the energy equation around which our economy is structured.

Photovoltaics becomes economic at different unit costs, depending on the application and relevant variables such as the climate, the tax status of the owner, the cost of money, and the price of utility-generated electricity. For a number of reasons, we use the figure of $3 per installed Watt of peak capacity as a benchmark at which photovoltaic arrays will become generally competitive for residential applications in the United States (all prices are in 1984 dollars).

PURIFYING SILICON

Most of the approaches to fabricating photovoltaic cells rely on silicon. Because a prime objective is to reduce costs, an adequate supply of low-cost silicon of the required purity is a basic concern. The best source of silicon is silica, which is silicon dioxide, found in nature throughout the United States as high-grade sand. Quartz rock is another form of silica. To produce silicon, silica is fused to a melt, and a controlled amount of carbon is added in a batch process. The carbon reacts to remove the oxygen as carbon dioxide, leaving relatively pure silicon, which cools into a block or ingot. This medium-gray, metallic-looking substance is metallurgical-grade silicon. Metallurgical-grade silicon has about 1 percent impurities and sells for $0.50–$1.00 per pound ($1.10–$2.20 per kilogram).

Converting this to semiconductor-grade silicon is expensive, as impurities must be cut greatly. The silicon must be extremely pure (except for the minute, controlled quantities of dopants) because ordinary impurities upset the delicate electron structure that exists between the atoms. In fact, the required purity of semiconductor-grade silicon is less than one extraneous atom per billion atoms of silicon. For comparison, in a world of some 4 billion persons, this is less than one "impure" person per continent.

The required purity is reflected in a selling price for semiconductor-grade silicon (in 1984) of $50–$70 per kilogram. At this price, the cost of silicon accounts for at least half of the cost of the finished cell. In addition, supplies of semiconductor-grade silicon are running low due to the expanding market for transistors and integrated circuits for computers, calculators, television sets, digital watches, and now the rising demand for photovoltaic systems.

Fortunately, photovoltaic devices do not have to use silicon as pure as semiconductor grade. They can tolerate ten times as much impurity as other semiconductor applications—as much as one millionth of one percent. New processes have been developed by American firms to make "solar-grade" silicon as well as semiconductor grade, and estimated prices of the product range from $10–$20 per kilogram. One or more new plants using these processes are expected to be in operation in the near future. The projected decline in the cost of photovoltaic systems is predicated in part on the availability of silicon of adequate purity at a lower price.

The Siemens process has been used for 20 years to provide semiconductor-grade silicon for the electronics industry. Very pure compounds of silicon are reacted with an electrically heated seed rod. Ultrapure silicon is slowly deposited on a cherry-red rod, which grows into an ingot weighing about 220 pounds (100 kilograms). This is a high-temperature, slow, batch process, and the product is expensive, primarily due to the large amount of heat energy consumed.

Though a solid block, the ingot is not a single crystal but a jumble of tiny crystals known as polycrystalline silicon, or polysilicon.

Base Option: Single-Crystal Silicon

The next step is to produce single-crystal silicon in block form. This is done by remelting the polysilicon in a special furnace where the temperature is carefully regulated. In the Czochralski process, a seed crystal of silicon on a holder is dipped into the melt and then very slowly drawn upward by

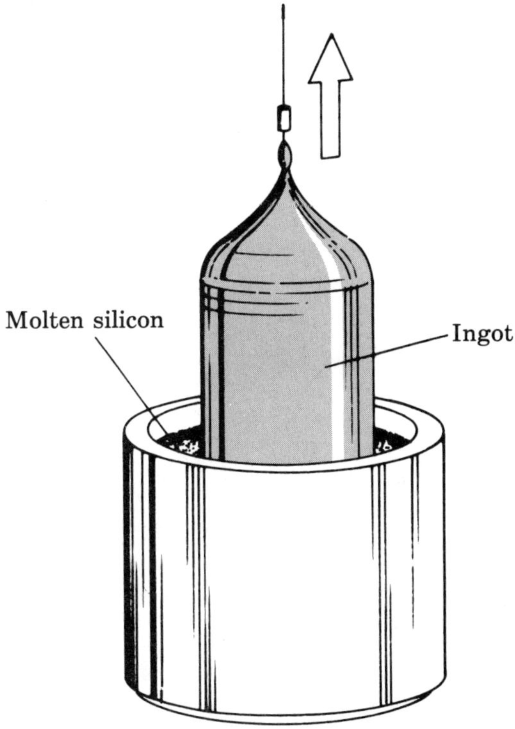

Figure 3-1 An ingot of single-crystal silicon being grown from a crucible of molten material (Czochralski process). The atoms of silicon solidify in perfect tetrahedral pattern following the structure of a seed crystal that is immersed in the melt, then slowly drawn upward. Under carefully controlled conditions, growth rates of as high as 4 inches (10 centimeters) per hour are possible. (Courtesy: U.S. Dept. of Energy)

mechanical control at a rate of perhaps an inch every three hours (Fig. 3-1). The molten material starts freezing (solidifying) on the seed crystal, following the identical crystalline pattern. In the most advanced production units available, single-crystal ingots can be formed 6 inches in diameter and up to 36 inches long. The price of crystal ingot is $100–$200 per pound. Monsanto Crystal Systems, Silicon Technology, Siltec, and Varian are leading producers of single-crystal ingots. To reduce costs, development effort is under way on continuously fed, higher-speed pulling operations.

After cooling, the cylinder is laid on its side and carefully sawed into thin sheets or wafers (Fig. 3-2). The object is to obtain as many wafers as possible from one bar while minimizing loss in the form of "kerf," or sawdust. Typical

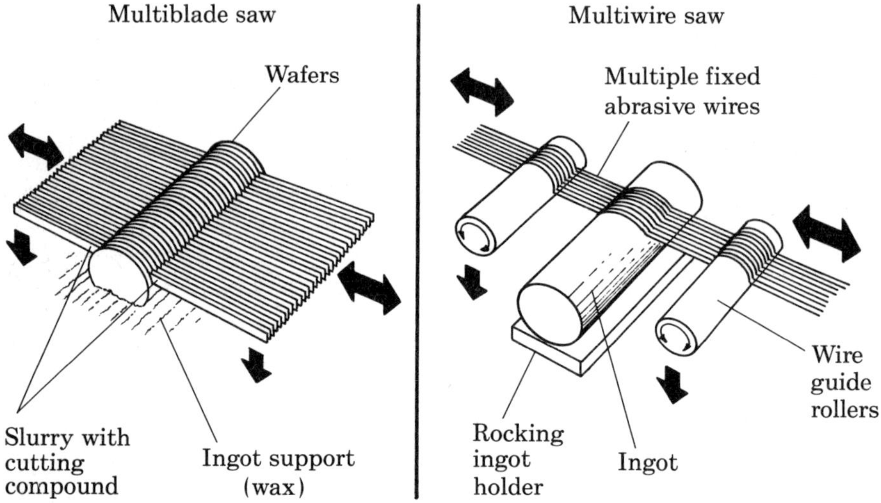

Figure 3-2 Methods of sawing silicon crystal into wafers. The cutting surfaces contain diamond abrasive. The wafers are 1/100 inch thick. Half of the expensive silicon bar is lost as dust. (Courtesy: U.S. Dept. of Energy)

wafer thickness is 1/100 inch (0.025 centimeter), while kerf losses are about 50 percent. Much has been done to improve sawing techinques and speeds. Thinner saws, multibladed circular saws, and wire saws of various descriptions have been developed, some of which can cut 1,000 wafers at a time. Companies prominent in wafering are Hamco, Siltec, and Varian.

Doping of the slice to create the two layers required for the photovoltaic phenomenon is accomplished in two steps. A trace of boron (or another element such as gallium) is first added to the molten silicon and thoroughly blended in. The ingot, as grown, is thus all p-silicon. After the wafers have been cut, they are coated on one side, then passed through a furnace containing a vapor of phosphorus (or another element with five outer shell electrons) for just enough time to allow the phosphorus to diffuse a short distance into the silicon, converting it into a very thin layer of n-silicon. At the zone of deepest penetration of the phosphorus, the barrier of static electricity spontaneously and immediately establishes itself. Metal contacts are then applied to each side—usually a metal plate to the bottom or p-side and a grid of metallic fingers to the top n-side. Finally, the entire cell is encapsulated in a weather-resistant transparent coating, such as a silicone plastic (Fig. 3-3).

If the pure silicon can be supplied for this base process at $15 per kilogam; if efficiencies are increased from 10 to 16–18 percent; if continuous, high-

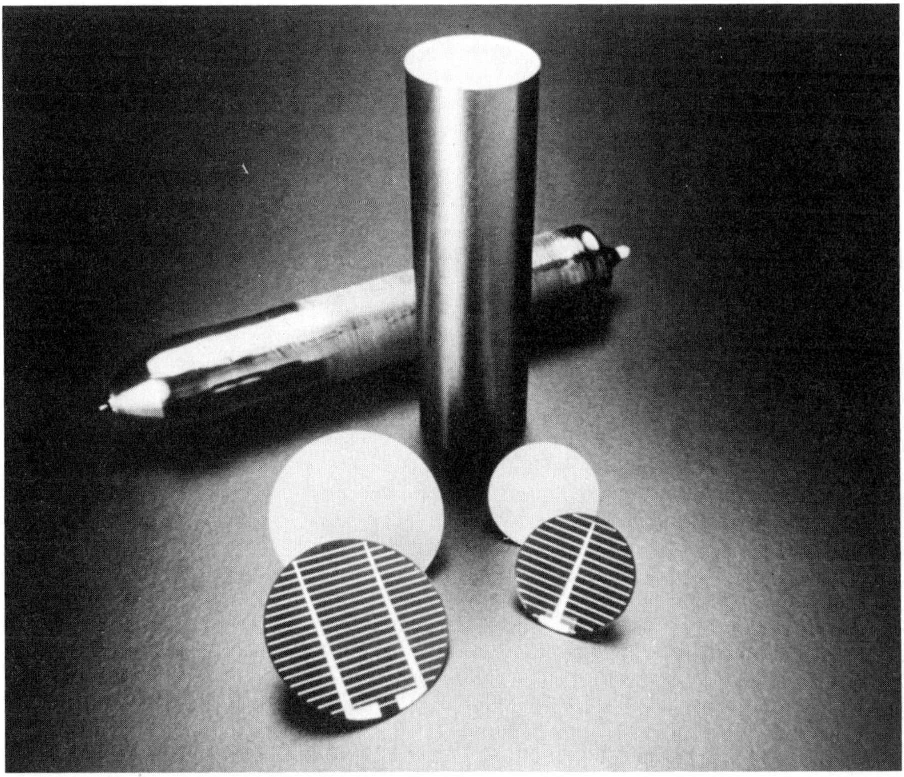

Figure 3-3 Czochralski ingots, wafers, and cells. (Courtesy: U.S. Dept. of Energy)

speed pullers are developed; if high-speed, multiple-blade slicing is achieved; and if automated production is implemented, single-crystal silicon arrays can be sold profitably at prices near $2 per peak Watt. At this array price, photovoltaic systems can be made and installed at costs of $2–$3 per peak Watt, generating electricity that is cost-competitive ($0.04–$0.12 per kilowatt hour) with alternative supplies in many areas. Over a hundred industry and university teams are researching these approaches to cost reduction.

Option 2: Silicon Sheet

To avoid the two expensive steps in making single-crystal wafers—pulling the bar and slicing it—methods have been developed to produce a sheet of silicon that is nearly single-crystalline in structure. There are two major sheet processes, the ribbon and dendrite processes.

In the ribbon process, a graphite or ceramic die, or shaping guide, that has a thin slit through the center, is lowered into a pool of molten silicon. The silicon is drawn up through the slit by capillary action. A seed crystal is touched to the melt at the top of the die and slowly withdrawn upward. As the silicon leaves the die, it slowly cools into pure crystalline form following the single-crystal pattern established by the seed. The melt is continuously replaced in this and similar sheet processes. This process is also called edge-defined film-fed growth (EFG). The product is a ribbon, 4–5 inches wide (10–13 centimeters) and less than 1⁄100 of an inch thick, which is wound onto a drum above the melt. Subsequently, it is cut into sheets of suitable length and doped and surfaced just as single-crystal wafers are. The resulting cell is rectangular (Fig. 3-4), preferable to the circular wafer.

The key companies involved in ribbon growth are Mobil Solar Energy Corp., Energy Materials Corp., and Motorola. The quality of the silicon produced by this method is adequate to produce 10–12 percent efficient rectangular cells that can be combined to make 8–10 percent efficient arrays. If one does a detailed cost analysis of this large-production volume, low-material-consuming option, it can be shown that array prices below $2 per peak Watt can be obtained.

The dendrite process developed by Westinghouse does not use a shaping guide. Very simply, two special seed crystals, called dendrites, are dipped side by side into molten silicon. The crystals are pulled out and up from the

Figure 3-4 A strip of crystalline silicon ribbon (below) grown by the EFG method. The ribbon is 3.9 inches (10 centimeters) wide and 0.012 inches (300 micrometers) thick. A cell (right) made from this material and a module (above) containing 12 cells. (Courtesy: U.S. Dept. of Energy)

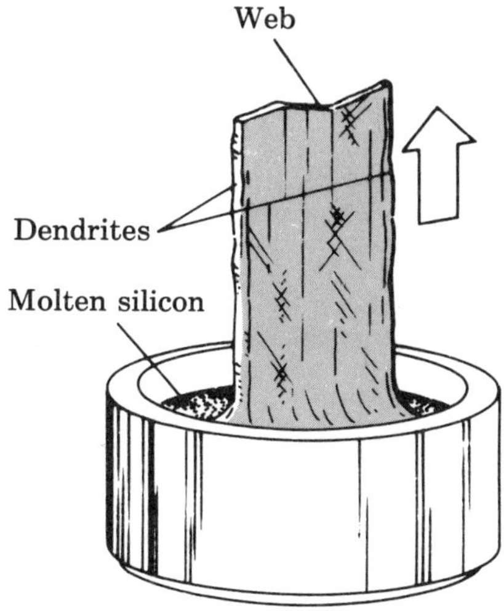

Figure 3-5 Web dendritic sheet growth. (Courtesy: U.S. Dept. of Energy, with Westinghouse, Inc.)

Figure 3-6 Future silicon sheet production. In this artist's conception of a large-scale production unit for silicon sheet, the material crystallizes as it is slowly withdrawn from molten baths. In practice, this would be done in an enclosed unit under an inert atmosphere and carefully controlled conditions. (Courtesy: U.S. Dept. of Energy)

melt, and a sheet of excellent-quality silicon crystal is grown continuously without slicing (Fig. 3-5). Solar cells made from dendritic web are excellent, with 16 percent efficiency the rule.

The technical feasibility of both the ribbon and the dendritic web processes has been established. Using experimental equipment, the pull rates, sheet width and thickness, quality, continuous processing and material consumption that are required to reduce the cost of arrays by the necessary factor of 10 have been demonstrated. It is possible that within two years, factories making continuous silicon sheet can be built (Fig. 3-6), assuring array costs and prices even lower than those forecast for the basic single-crystal sliced product.

Option 3: Ingot Casting of Nearly Single Crystals (Polycrystalline Silicon)

Single-crystal pulling can be replaced by a casting process in which pure molten silicon at about 2,000°F (1,100°C) is poured into a special pot or crucible and allowed to solidify in a controlled manner. The resulting ingot is composed of individual crystallites that are quite large (Fig. 3-7). The ingot is sliced using the advanced slicing technology of the Base Option. This method, called SEMIX (for semicrystalline material), has been pioneered on a proprietary basis by a subsidiary of Solarex Corporation. A German company, Wacker Chemie, is also pursuing this approach.

Alternatively, the molten silicon is allowed to "freeze" very slowly above a seed crystal that is artificially cooled. The product is a block of true single-crystal material, which subsequently is sliced. Known as the heat exchange method (HEM), this process is under development by Crystal Systems, Inc.

The two processes are compared in Table 3-1. The SEMIX process is faster and uses much less energy than the HEM process. Both processes consume smaller amounts of pure silicon than the Czochralski method. The SEMIX process may be able to use silicon of much lower purity than the HEM process. Single-crystal advocates assert that semicrystalline material will be harder to saw because of chipping, and they say that cell efficiences will be much less than from the single-crystal HEM material. Recently, the HEM process has been modified to produce a large-grain polysilicon similar to SEMIX. The market will decide which approach is viable.

Products from both processes were sold as arrays in 1984. Cost analyses show that array price goals of less than $2 per peak Watt can, indeed, be met by either method if pure silicon is available at $15 per kilogram. If these claims are substantiated, the requisite low array prices will clearly be achievable.

Figure 3-7 Square ingot by the SEMIX method. More than 6 inches on a side, this solid block of polycrystalline silicon weighing 21 pounds (9.5 kilograms) was produced by slow cooling of molten material. Crystallite faces are clearly visible in wafer at the right. After sawing, the wafers will be made into cells and assembled into a compact module. The advantage of the rectangular shape is that the wafers make square cells that can be fitted closely together in a module, saving space without wasting silicon. (Courtesy: U.S. Dept. of Energy)

Table 3-1 HEM vs. SEMIX

	HEM	*SEMIX*
Feasibility	Yes	Yes
Material	Single crystal or multicrystal	Multicrystal
Device Quality	Excellent	Excellent
Cost	Less than crystal-pulling	Much less than pulling
Factory Status	Experimental	New proprietary factory

SOME LONGER SHOTS: MATERIALS AND DESIGNS

Fortunately, but adding to the complexity of the situation for those who must orchestrate the development of photovoltaics, there are several other approaches to designing and producing photovoltaic cells that are equally exciting from the standpoint of reducing costs. In fact, the next three options all appear capable of providing modules that cost less than $2 per peak Watt.

Option 4: Non-Silicon Thin Films

This category encompasses thin films made with cadmium sulfide, copper sulfide, zinc sulfide, cadmium telluride, and similar compounds. The elegance of devices made from these materials lies in simplicity of manufacture. Some of the materials can be simply spray-coated on glass, dipped in copper, and processed at temperatures less than 400°F (200°C) in a continuous fashion. Because the photosensitive layers are ultrathin, very little material is consumed, and cell costs of less than $0.50 per peak Watt will be possible.

The University of Delaware, in conjunction with DuPont, has developed a detailed plant design, obtaining costs and selling prices much less than $2 per peak Watt. Figure 3–8 shows an artist's concept of a fully continuous cadmium sulfide production process, which is not too unlike the production of paper. Photon Power in El Paso, Texas, ran a pilot line making 2 foot by 2 foot (61 centimeter by 61 centimeter) arrays. Even at efficiences as low as 4–6 percent, these cells would cost less than $1 per peak Watt. One might ask, then, "If this technology is so inexpensive, why isn't it widely marketed now?" There are two reasons. The low relative efficiency requires the use of larger array areas, increasing area-related costs such as mountings. In addition, the lifetimes of the arrays have yet to be demonstrated. SES, a subsidiary of Shell Oil Company, made arrays for nearly two years and tested them thoroughly. Shell management delayed extensive private sales of their cadmium sulfide arrays because the lifetimes were considered too uncertain for the corporation to put its name on the product. In 1983, Shell dropped the cadmium sulfide line and joined Motorola in developing the ribbon-to-ribbon process.

The University of Delaware, in a more sophisticated, more costly process, has demonstrated over 10 percent efficiency on small cells in the laboratory. Again, the key issues are the lifetime and efficiency of cells produced in a continuous, low-cost fabrication process. However, the general view is that the cadmium zinc sulfide compounds may offer a practical route to very low-cost arrays. The reliability of these materials will in time be proven and penetration into the market can be expected. This intrinsically simple manufacturing approach may eventually present strong competition to all-silicon technology, at least for some applications.

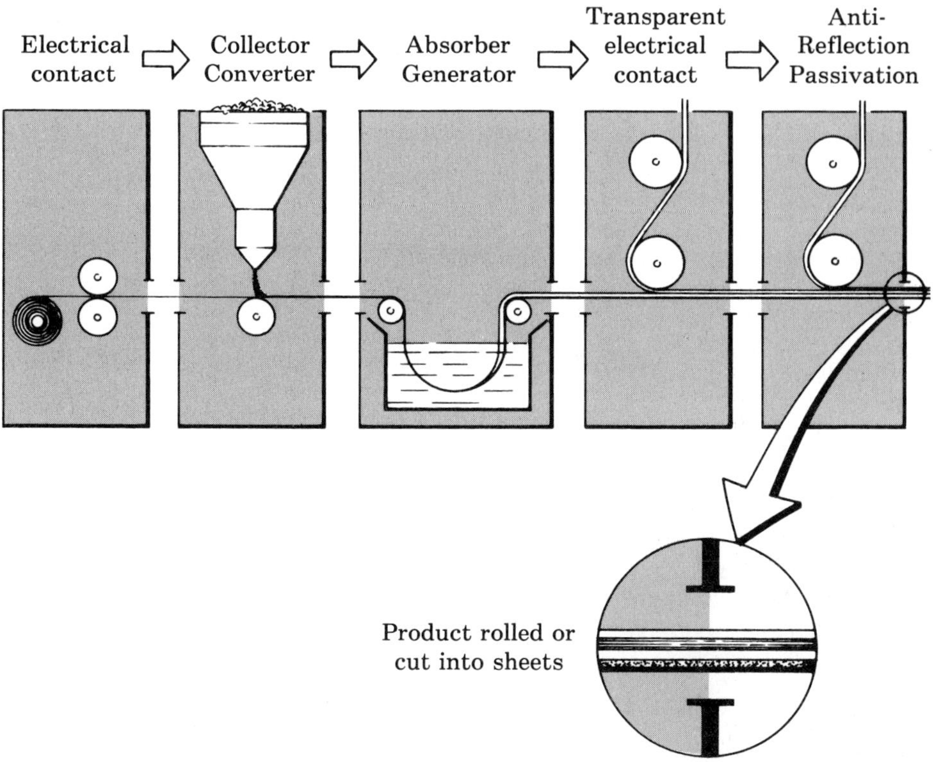

Figure 3-8 Cadmium sulfide/copper sulfide continuous production process. The schematic diagram above shows how a solar cell might be produced in a continuous process. A roll of thin copper sheet (left) is drawn into the second stage, where a thin deposit of cadmium sulfide is added. It then passes through a dip of copper sulfide solution (center), where a thin film plates out to form the cell junction. Next, a thin film of tin oxide, which serves as the upper electrical contact, is added. Finally (right), an anti-reflective coating and weather surface completes the cell structure, which has a total thickness of 0.002 inch (50 micrometers). The material has then only to be cut into convenient lengths and electrical connections added. (Courtesy: U.S. Dept. of Energy)

Option 5: Concentrator Designs

While the photovoltaic cells described so far are mounted flat-plate systems that use sunlight without magnification, another option uses optical concentration systems of lenses or mirrors to focus sunlight on very high-efficiency cells made of single-crystal silicon or gallium arsenide (Fig.

3-9). The cells get hot, which is a disadvantage because cell efficiences decline at higher temperatures. However, this disadvantage can be partially offset if the heat is used in addition to the electricity.

Using concentrators made of plastic, glass, and aluminum decreases the need for more expensive semiconductor material. If the sunlight is concentrated 100 times, 1/100 of the expensive semiconductor solar cells is required. The high efficiency of the clever designs developed by the concentrator

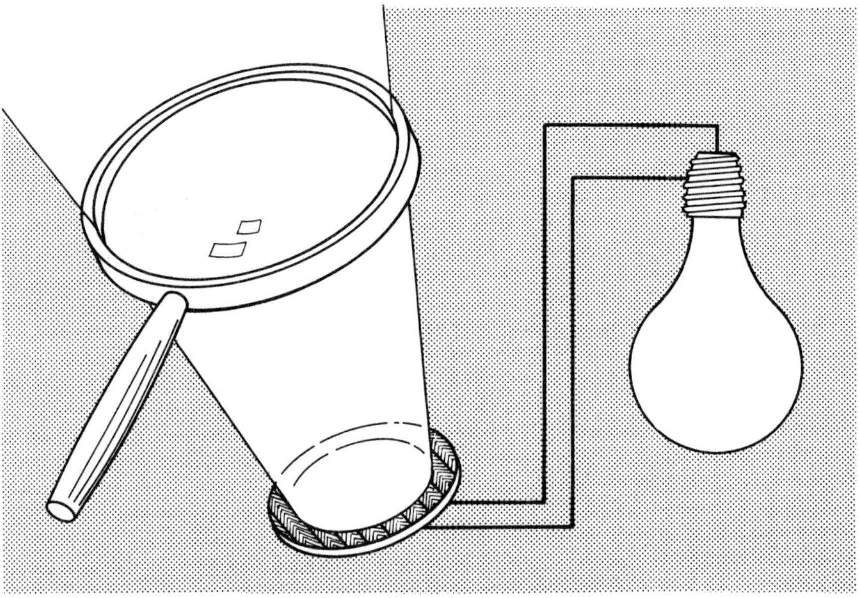

Figure 3-9 Concentrator cell. By focusing sunlight on a small area, more expensive cells can be used, often with increased overall efficiency. The disadvantages are that the entire unit must track the sun and diffuse light is not utilized. Concentrator systems are economic primarily in areas that have a large number of clear days during the year. (Courtesy: U.S. Dept. of Energy)

Figure 3-10 Three leading methods of concentrating sunlight on photovoltaic cells: (a) Parabolic trough reflector focuses sunlight on a string of cells as indicated in the cross-sectional diagram. The reflector is rotated to follow the sun; (b) Linear Fresnel lens arrangement. The lens focuses sunlight on the row of cells just as a large magnifying glass would do. The whole device turns to follow the sun. (c) Point-focus Fresnel lens concentrates sunlight on a single cell. This arrangement requires two-axis tracking in order to follow the sun exactly. All require a coolant, usually a liquid pumped across the back side of the cells. (Courtesy: U.S. Dept. of Energy)

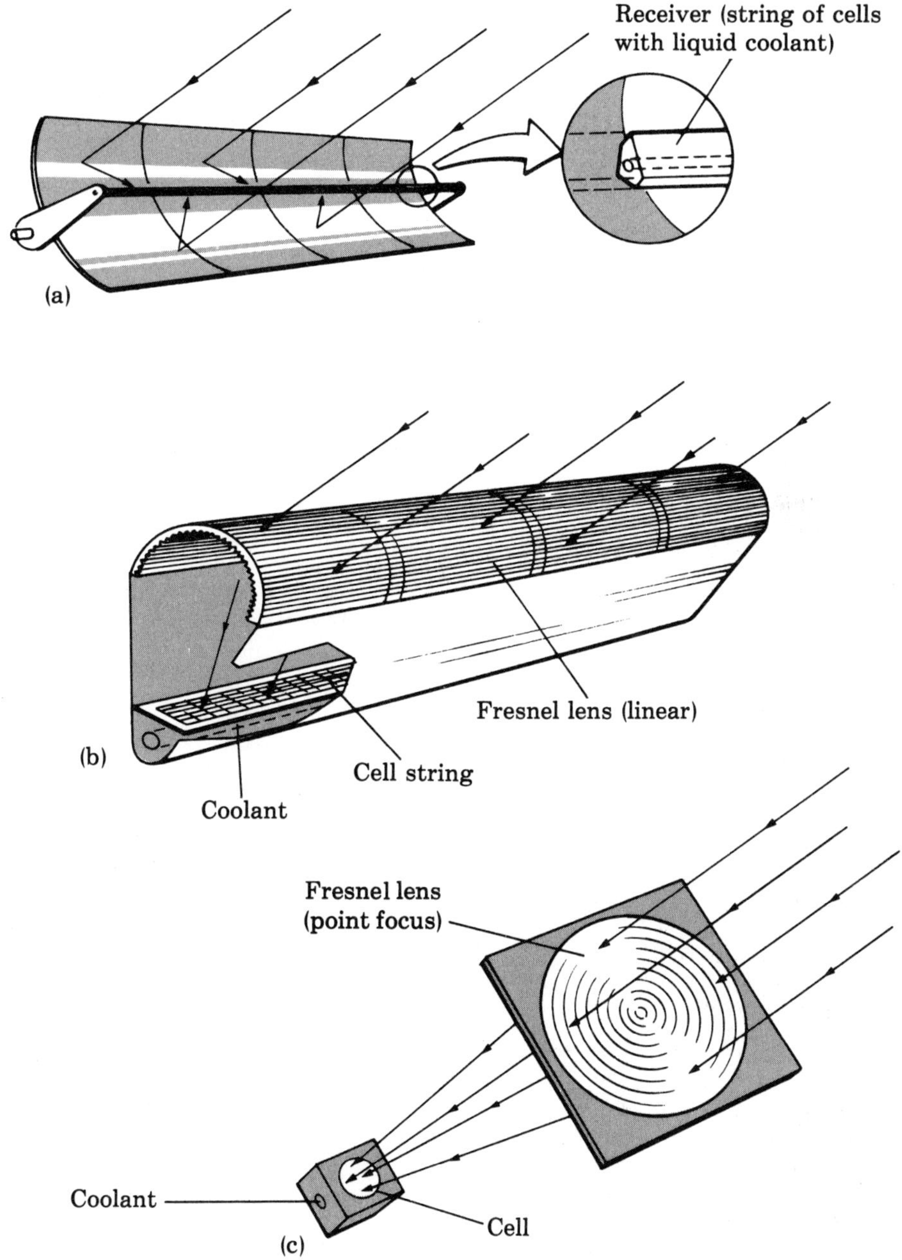
Receiver (string of cells with liquid coolant)
(a)
Fresnel lens (linear)
(b)
Cell string
Coolant
Fresnel lens (point focus)
Coolant
Cell
(c)

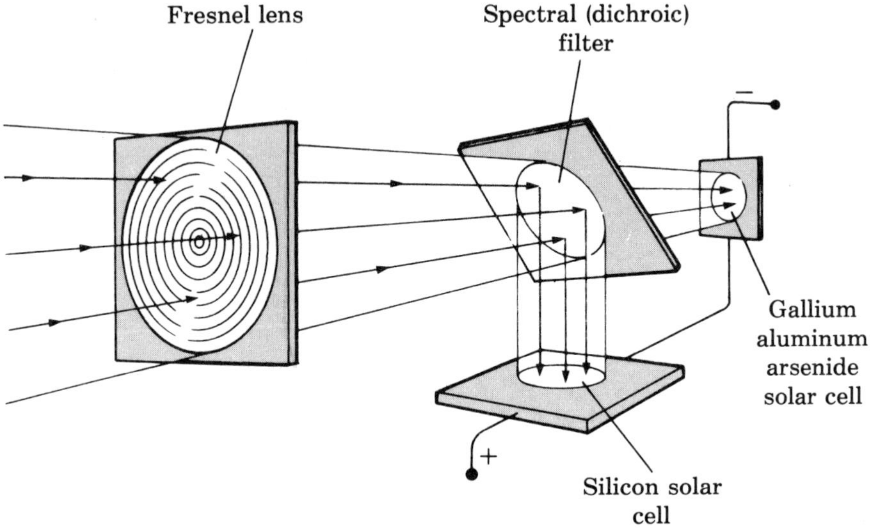

Figure 3-11 Split-spectrum cell. Concentrated sunlight is passed through a special filter that reflects the red part of the spectrum to a silicon cell and allows the blue part to pass through to activate a gallium arsenide cell. Because each cell is more responsive to the light it receives, the overall conversion efficiency is increased. (Courtesy: U.S. Dept. of Energy)

industry have led to forecasts from the private sector for selling prices as low as $1 per peak Watt. Typical optical configurations for concentrating light are shown in Figure 3-10. The Department of Energy (DOE) awarded contracts for five large concentrator experiments to supplement its $10 million concentrator cost reduction effort. Using expensive compound semiconductor cells, Varian Associates reported efficiencies as high as 33 percent for cells and 25 percent for the entire system.

Four basic approaches are being tried to increase electrical conversion efficiencies of photovoltaic concentrators. The first is to make the standard single-silicon and gallium arsenide solar cells themselves more efficient. Progress by Microwave Associates indicates that silicon concentrators using sunlight concentrated as much as 1,000 times can convert at 22–24 percent efficiency, based on the incoming sunlight. United Energy Corporation installed 4½ megawatts of Fresnel lens systems, having 13 percent efficiency, in 1983. Intersol shipped a 20-kilowatt point-focus Frȇsnel lens concentrator with over 15 percent efficiency in 1984.

The second approach is to split the solar beam into two or more parts of different wavelengths (i.e., different colors), each of which is then focused

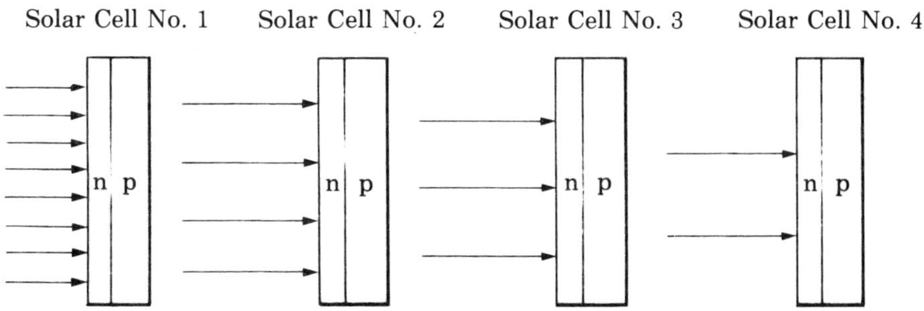

Figure 3-12 Multiple-junction tandem cell. In this system, three or more cells of different materials are stacked one on top of the other. The top cell absorbs and is activated by the higher-energy photons; the next cell down converts photons of lesser energy; and the bottom device absorbs still longer wavelengths. Even though more complex, such systems may in time become economically practical when very thin films are in mass production. (Courtesy: U.S. Dept. of Energy)

on a separate photovoltaic cell most sensitive to that part of the spectrum. At Varian Associates, laboratory experiments employing dual silicon and gallium arsenide cells have achieved over 30 percent efficiency (Fig. 3-11). There is no reason why this elegant use of two or three cells cannot lead to manufactured concentrator systems that convert 25 percent of the solar energy into electricity and the remaining 75 percent into heat at 212°F (100°C). Such a system is well suited to the industrial electricity and heat market.

In the third high-efficiency approach, semiconductors having different frequency response to the solar spectrum are stacked on top of each other to make a composite collector. These cells are still under development, but theoretical calculations indicate conversion efficiencies as high as 60 percent are possible. In this multiple-junction or tandem-cell system, sunlight passes successively through three thin cells of different materials lying one above the other (Fig. 3-12). The top cell absorbs and is activated by the high-energy photons, the next cell converts photons of lesser energy, and the bottom device absorbs still longer wavelengths. Efficiencies of 20–25 percent have been achieved. Even though more complex, such systems may in time become economically practical when very thin films are in mass production.

The fourth high-efficiency approach is called thermal photovoltaics (TPV). This process, being developed at Stanford University, uses concentrated solar energy to heat a radiator surface to a very high temperature. This radiator then emits the energy in a different part of the spectrum, which better matches the silicon target cell. This concept is depicted in Figure

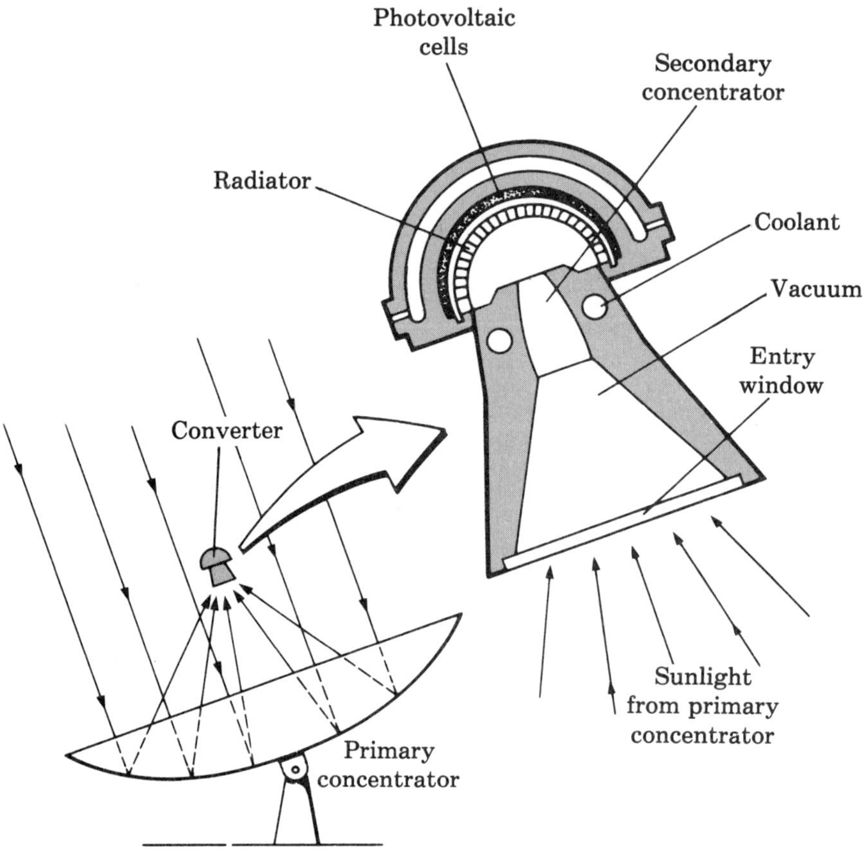

Figure 3-13 Thermophotovoltaic cell. Sunlight is focused on a metal plate, heating it red hot. The plate then reradiates the heat at a longer wavelength, and the latter activates the photovoltaic cell. The silicon cell is more sensitive to the longer wavelength, increasing overall conversion efficiency. (Courtesy: Electric Power Research Institute)

3-13. The silicon cell converts the reradiated energy to electricity. Calculations by SAI Corp. for the Electric Power Research Institute indicate 30–35 percent overall conversion efficiencies can be obtained using the TPV process.

It seems clear that high-efficiency photovoltaic concentrators will find a niche in the marketplace. The wide range of concentrator configurations, cell approaches, and materials of fabrication all forecast a period of intense competition, as well as some confusion, as the various options sort themselves out in the marketplace.

Option 6: Amorphous Materials

The ultimate solution for photovoltaics could well be the use of amorphous materials deposited on steel or plastic. Very little semiconductor material is used, expensive crystallization procedures are bypassed, and the amorphous state is very durable. Presently, 6 percent efficient amorphous silicon films doped with hydrogen are being produced by Sanyo and Fuji in Japan. Other developers have taken the amorphous silicon and added fluorine instead of hydrogen to make improved material.

Energy Conversion Devices, Inc. (ECD) has shown in theory that fluorine-silicon thin film can provide 10 percent efficiencies at quite low cost. ECD is likely to make cells with higher efficiences than the 4–7 percent efficient cells being sold today. If amorphous material can provide cell efficiencies of 6 percent with cell lifetimes of ten years, there is no doubt that selling prices of less than $1.50 per peak Watt will be forthcoming. Both RCA and ECD have produced experimental cells of about 1 square centimeter with efficiencies greater than 10 percent. We forecast inexpensive commercial cells of 8 percent efficiency within three years.

Chronar Corporation is developing a chemical vapor deposition process for making amorphous silicon cells. In 1982, Chronar achieved efficiencies of 6.5 percent on cells as large as 5 centimeters by 5 centimeters. Production began late in 1983. Chronar constructed three plants in 1984. Arco Solar tested 7–8 percent panels in the same year. Solarex purchased rights and equipment from RCA in 1983 and shipped calculator cells in early 1984.

Several million dollars of risk capital from U.S. industry, in addition to $5 million per year in federal contracts, have been committed to the investigation of amorphous silicon. The Japanese are also concentrating on amorphous silicon. Sanyo has a complete line of watches, clocks, and calculators powered by amorphous silicon. Sanyo discovered that amorphous silicon will convert artificial fluorescent light at 6–8 percent efficiency. The low cost of amorphous material coupled with efficiencies close to that of single-crystal silicon cells led Sanyo to commercialize amorphous silicon. This may have been a decisive event in the history of photovoltaics.

EXPERIMENTAL CELL TYPES

There are two special experimental cell types that deserve brief mention. The vertical multiple-junction cell has tiny grooves cut across the back. Each

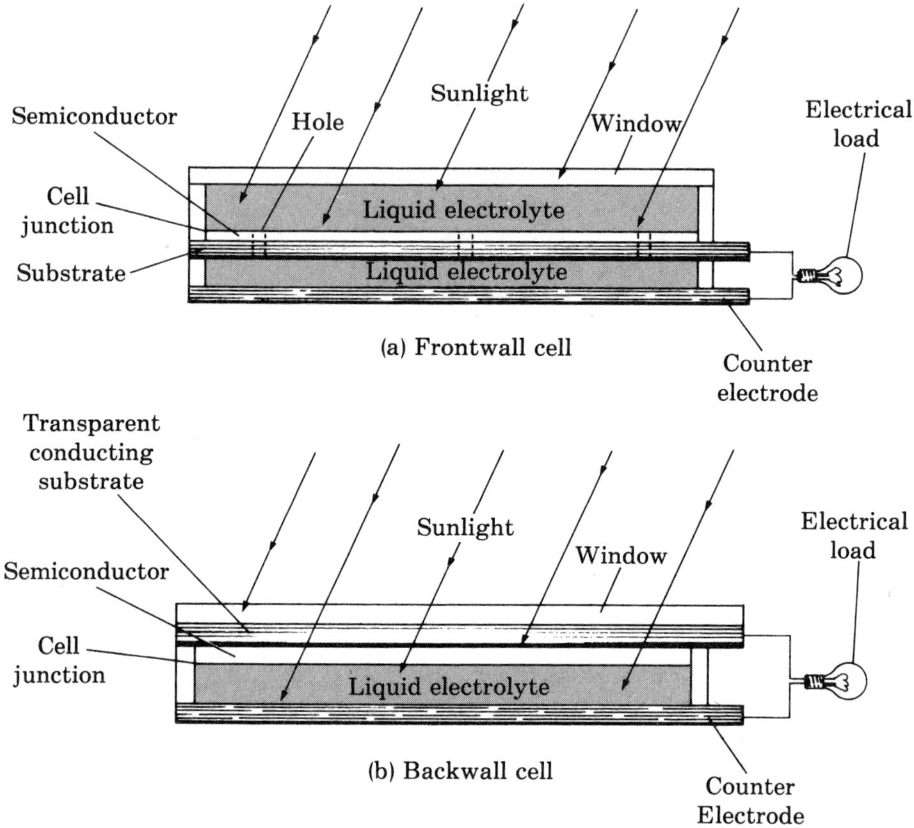

Figure 3-14 Electrochemical photovoltaic cells. The remarkable versatility of photovoltaics is illustrated by the electrochemical photovoltaic cell, which employs a liquid phase. The cell junction is established at the interface between an electrolyte solution and solid semiconductor material. It was in a simple battery containing a liquid electrolyte that the photovoltaic phenomenon was first observed by Becquerel in 1839. Two designs are shown in which the electrolyte is (a) above and (b) below the cell junction. (Courtesy: U.S. Dept. of Energy)

groove is coated on the inside to form the cell junction, greatly enlarging the barrier area and thus increasing cell efficiency. Conversion efficiencies of about 20 percent have been achieved with this unusual design.

Another unique type is the electrochemical cell (Fig. 3-14). Its special feature is that one layer is liquid. Because the cell junction is formed at the interface between a solution of an electrolyte and a solid semiconductor, some aspects of fabrication are simplified. This cell opens the door to

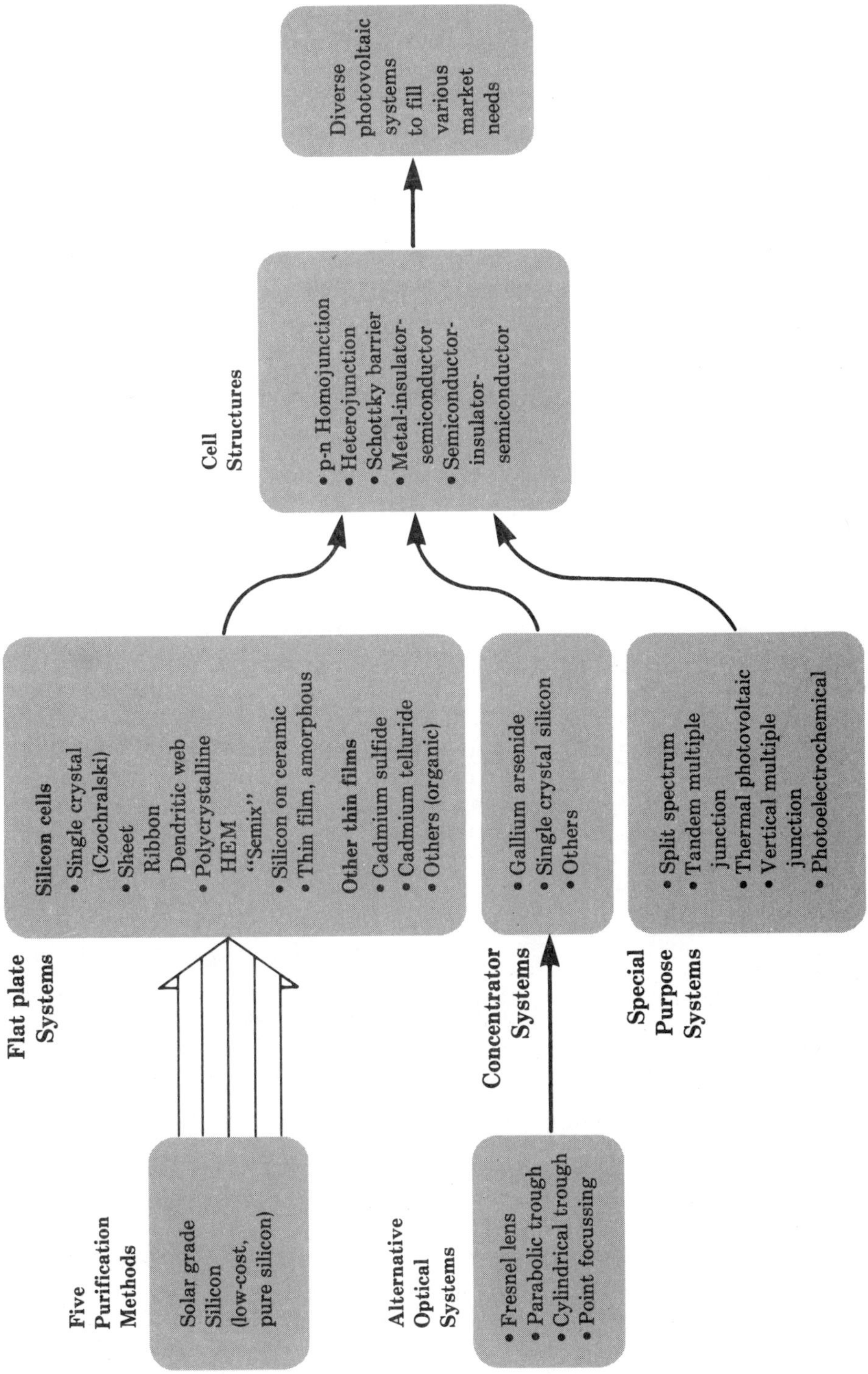

Figure 3-15 Diversity of approaches to meet various market needs at reduced cost. (The silicon-on-ceramic option is not currently being pursued.)

unique applications. For example, the electric energy generated by the cell can be withdrawn for external use as with any other photovoltaic cell, or it may be used within the cell itself to produce useful by-products. In the latter case, materials rather than electricity are withdrawn. If hydrogen is produced, for example, it can be piped off and stored for later use when the sun is not shining. Research now under way promises an electrochemical cell that is 10 percent efficient in converting sunlight into energy. Current research is focusing on several semiconductor materials in single-crystal, semicrystalline, and amorphous forms for use in electrochemical cells, and on the effects of using different kinds of electrolytes: aqueous, nonaqueous, molten salt, and even a solid electrolyte.

A TECHNOLOGY TEEMING WITH POSSIBILITIES

Photovoltaics is a rich and promising technology. The basic photovoltaic phenomenon of converting sunlight to electricity encompasses a diversity of materials and a multiplicity of fabrication techniques, and these can be coupled with a wide assortment of cell designs.

To this is added the flexibility of using either flat-plate or concentrating optical systems, as well as the potential of concepts such as multiple-junction, split-spectrum, and thermophotovoltaic cells. Many approaches are available to reduce costs through simplifying and automating production processes. Simultaneously, scientists are learning more about the delicate effects of molecular structure and trace materials on conversion efficiencies. The multiple avenues that are open assure continuing rapid decline in system prices even if there are no further breakthrough discoveries.

The production technology, which is evolving hand-in-hand with advances in cell design, is equally rich in opportunities for discovery and invention. There is, to be sure, an irony in this state of affairs, for the very diversity and promise of the technology diffuses available venture capital among a host of candidate solutions, no one of which is clearly superior or preeminent to all others.

FURTHER READING

Ariotedjo, A. P., and Charles, H. K., Jr. "A Review of Amorphous and Polycrystalline Thin Film Silicon Solar Cell Performance Parameters." *Solar Energy* 24, November 1979, pp. 329–339.

Backus, C. E. "Photovoltaics III: Concentrators." *IEEE Spectrum* 17, February 1980, pp. 34–36.

Backus, C. E., ed. *Solar Cells.* New York: IEEE Press, 1976.

Basic Photovoltaic Principles and Methods. Solar Energy Research Institute. SERI/SP-290-1448. Golden, Colo.: Solar Energy Research Institute, 1982.

Carlson, D. E. "Photovoltaics V: Amorphous Silicon Cells." *IEEE Spectrum* 17, February 1980, pp. 39–41.

Gay, Charles F. "Solar Cell Technology: An Assessment of the State of the Art." *Solar Engineering* 5, March 1980, p. 15.

Loferski, J. J. "Photovoltaics I: Solar-Cell Arrays." *IEEE Spectrum* 17, February 1980, pp. 26–31.

Loferski, J. J. "Photovoltaics IV: Advanced Materials." *IEEE Spectrum* 17, February 1980, pp. 37–38.

Maycock, Paul D.; Shimada, K.; Stirewalt, E.; and Hunt, D. *America Challenged: Photovoltaics in Japan.* Alexandria, Va.: Photovoltaic Energy Systems, Inc., 1982.

Wolf, Martin. "Photovoltaics II: Flat Panels." *IEEE Spectrum* 17, February 1980, pp. 32–33.

CHAPTER 4

PRESENT AND FUTURE APPLICATIONS OF PHOTOVOLTAIC ENERGY SYSTEMS

History is responsible for the mistaken and widely held notion that photovoltaics is a high-cost, exotic technology with very limited and specialized uses. In the rush to loft the American response to the Soviet Union's Sputnik, Army Signal Corps engineers at Fort Monmouth, New Jersey, obtained enough design latitude to power the Vanguard I's 5-milliwatt radio transmitter with commercially available solar cells. On March 17, 1958, six small arrays containing 108 silicon chips went into space.[1] But the full implications of using solar power did not sink in until after the launch: With no cutoff device, Vanguard's signals fully and needlessly occupied a radio band for about eight years. The next 22 U.S. satellites went up with electrochemical batteries, but in late 1959 Explorer VI was fitted with 8,000 1 centimeter by 2 centimeter silicon cells, which produced about 15 Watts of power.

Except for a few nuclear reactors, silicon solar cells remained the chosen power source in space. By 1975, the National Aeronautics and Space Administration (NASA) was using nearly a million cells a year. Array sizes grew: Nimbus, a weather satellite launched in August 1964, carried a 500-Watt array; the Orbiting Astronomical Observatory carried a 1,000-Watt array; some Air Force satellites produced 1.5 kilowatts. Only engineering considerations of balance, rotation, and stress now seem to present limits to the size of photovoltaic arrays in space. A manned space station plan calls for a 100-kilowatt array.

On earth, solar cell production grew rapidly during the early 1960s. Companies such as Hoffman (now Applied Solar Energy), Heliotek, International Rectifier, RCA, and Texas Instruments entered the field. By 1970, only two of these firms remained in the photovoltaics business. Total sales leveled off at about 80 kilowatts per year, at an average cost of $100–$200 per Watt.

Success in space led to renewed speculation among photovoltaic producers over potential uses on the earth. By 1970, concerns about finite fossil fuel supplies and about pollution from excess heat, combustion by-products, and radiation from other energy sources began to place photovoltaic power generation in an ever more favorable light.

In 1971, a NASA employee, the late William Cherry, pointed out that solar arrays covering 31,500 square miles (about 1 percent of the land area in the 48 contiguous states) could supply all the U.S. electric needs then projected for 1990. If the installed price could be reduced to $0.50 per Watt, and if the cells were to operate at no better than 7 percent efficiency, farmers could install arrays on otherwise unused land and earn perhaps $2,000 per acre per year from the sale of electric power. Operating costs over a 20-year period would total slightly more than half the costs for an equivalent hydroelectric plant, the next cheapest alternative.

Around 1973, three new manufacturers entered the field with an eye toward a new market for applications on earth. Today, about 18 companies are engaged in the manufacture of solar cells, equipment, and arrays in the United States. But all face the common problem of transforming a market perceived as small and specialty-oriented into one characterized by large scale, low unit cost, and widespread general applications.

At the same time that photovoltaics suffers from the misperception that it is an exotic technology with specialized applications only, experience in space and on earth has shown that this technology is well-suited for widespread use. Solar cells have survived the harsh physical rigors of space—high vacuum, hard radiation, great temperature differentials—and proven themselves durable, efficient, and dependable. In applications on the earth, they have survived dirt, dampness, chemical smog, wind, and hail, and again proven their virtues. Experience has shown that photovoltaic systems can survive in space for 20 years, and that they can survive the earth's environment (in many respects even harsher) for at least 10 years without signs of fundamental failure. Terrestrial systems have already been built and used for virtually every electrical load, ranging from microwatt systems for digital watches to remote systems powering houses, villages, and utility substations.

Assuming the ability to produce cheap photovoltaic cells, where and how are these millions and billions of tiny generating units going to be aggregated and used? There are three broad categories for photovoltaic applications on the earth: stand-alone, residential and other buildings, and central utility generation. Each of these must meet certain common design criteria, yet each has different operating criteria and economics.

One set of design criteria centers on photovoltaics' ability to tolerate the insults heaped on installations by the earth's weather. Many installations over many years have shown that weather, except to the extent that it limits insolation, is not an insurmountable obstacle to photovoltaic deployment in most areas of the world.

It is obvious that outdoor photovoltaic installations must be protected against moisture. This is a simple design problem. Beyond this, any precipitation that covers or condenses on the array surfaces—rain, sleet, snow, fog—can influence how well a solar array performs. Freezing rain or frost can cause some loss of power until it melts or evaporates with rising temperatures. Worse is snowfall followed by several days of subfreezing temperatures. Until most of the snow has melted from the array surface, little or no power can be generated. Experience with installations at General Electric's Valley Forge, Pennsylvania, office building and the Beverly, Massachusetts, High School, which endure both inclement and variable winter weather, has shown that neither snow cover nor frost has been a significant problem.

If any portion of the array is exposed to sunlight, even an edge or corner of a panel, the heat generated in that small part spreads across the array, which clears itself in a short time. The snow cover slides off the tilted surface, provided there is room at the bottom of the panel. In short, we are not facing a limitation on photovoltaic installations imposed by weather patterns, but a straightforward design problem.

Because ambient air temperature can affect cell temperatures, it affects the efficiency of the photovoltaic generators. The power output of a photovoltaic cell at a given level of insolation is inversely proportional to the cell temperature. For silicon cells, the power output falls about 10 percent for each 45°F (25°C) rise in cell temperature.

While the photovoltaic cell is the heart of an installation, it is not the only necessary part (Fig. 4-1). Solar energy (sunshine) is diffuse. The problem this presents to the system designer is twofold: the system developed must be as efficient as possible in collecting and converting the energy into usable form, and the collectors must cover large areas at minimum cost.

A photovoltaic system must have a structure to mount the cells on and wires to interconnect them and carry off the electricity (Fig. 4-2). Other necessary components of the system depend on how the system is to be used. If power is required when the sun is not shining, a battery for storage or another source of electricity must be included. When storage is used, a special kind of switch is added to prevent electricity from flowing from the battery back into the cells when they are not producing.

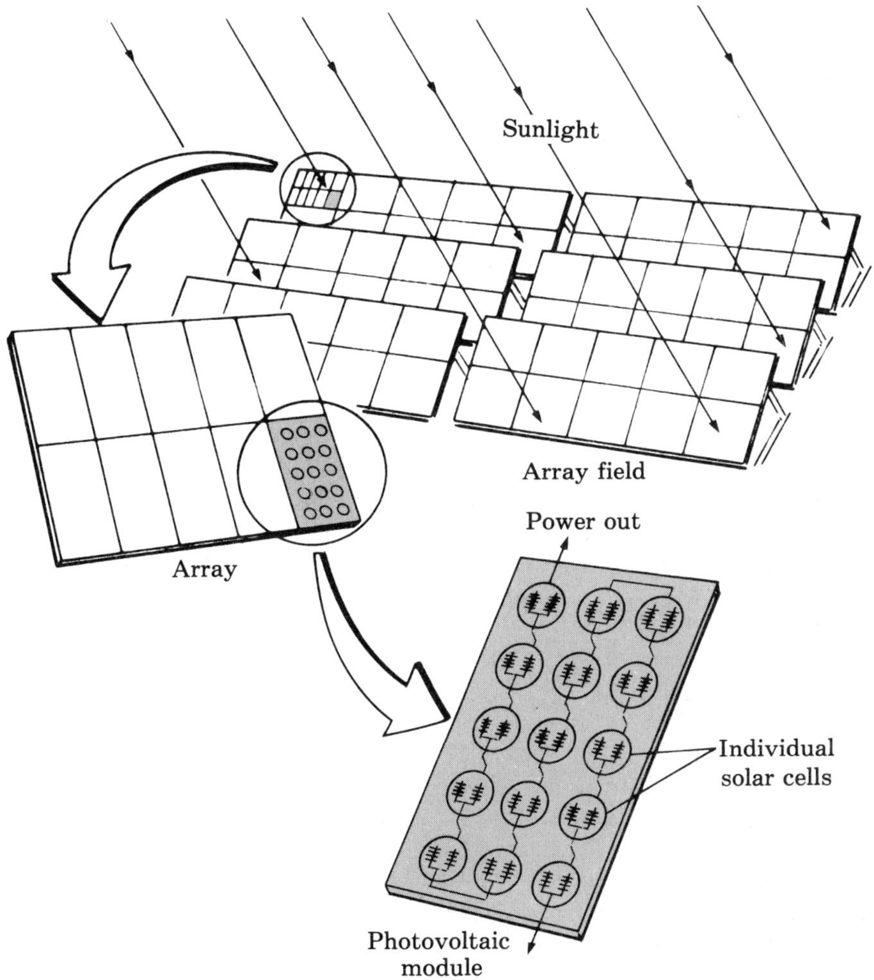

Figure 4-1 Module. The basic building block for photovoltaic arrays is the module, a sealed panel containing a number of cells interconnected at the factory. It is shipped as a unit, with electrical leads attached for connection to other panels or to an external circuit. Several modules are usually mounted together, forming an array. A collection of arrays is called a field.

Power produced by any photovoltaic device is direct current (DC), as the electrons are flowing steadily in one direction only. DC power can be used directly in some cases, but if the electricity is for household use, alternating current (AC) is required for most appliances. To change DC to AC, a power inverter is added. A photovoltaic system may include additional controls

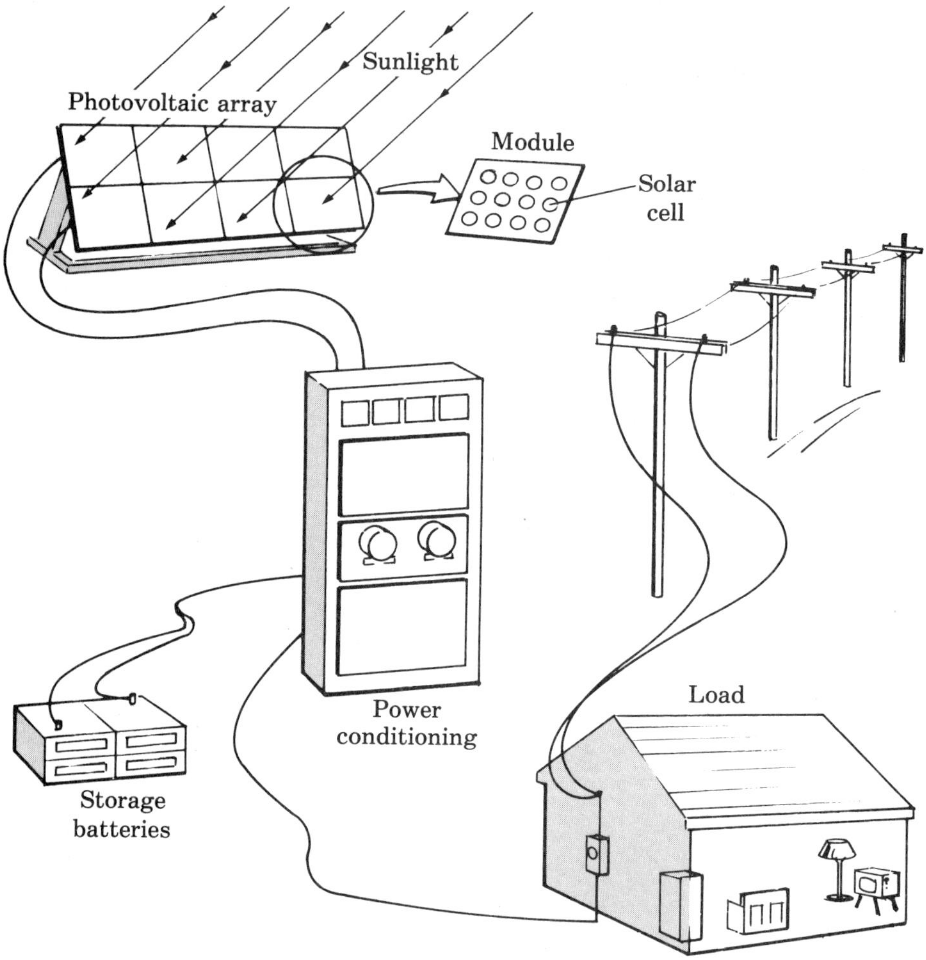

Figure 4-2 System components. In addition to the array, a photovoltaic system has a power conditioning unit, possibly storage batteries, and a connection to the using equipment or load. It may or may not be interconnected to a utility through a meter.

The components other than the array are sometimes referred to as the balance of system, or BOS. There is nothing exotic about these components. They can be assembled into systems with a minimum of engineering development. In mass production, system costs will decline significantly.

to monitor system performance and to regulate power being sold back to the utility.

If the system "stands alone," for example to power an isolated village that has no utility service, automatic controls may be incorporated to turn off power to less important equipment in the evening or during a period of cloudy weather. This controlled load shedding assures that power will be available where and when it is most needed without oversizing the system at unnecessary additional expense.

Collectively, components other than the cells are termed the balance of system (BOS). While they are a necessary part of the total photovoltaic system, there is nothing exotic about these components. They can be assembled into systems with a minimum of engineering development. They do have associated costs, which must be factored into evaluation of a particular installation, and the state of development of BOS equipment, such as batteries, can affect the capabilities of the system as a whole. Design criteria for the BOS components are, generally speaking, straightforward problems in engineering and are important primarily because of their impact on system costs and performance.

STAND-ALONE APPLICATIONS

As the name suggests, these photovoltaic installations must be virtually self-sufficient. They are not hooked into the utility grid, may have diesel, battery, or other backup systems, and are not regularly and conveniently supplied with fuel or maintenance. With some notable exceptions, e.g., low-lift pumping, stand-alone applications require that the photovoltaic system generate enough electricity from the sun to meet reliably the demand of the load for 24 hours per day, 365 days per year. Most stand-alone systems involve a photovoltaic array, battery storage, minor numbers of controls, and the load. Rather than having the battery replaced at regular intervals, stand-alone photovoltaic systems supply current to a rechargeable battery. Presently, worldwide, several thousand such systems are purchased each year.

Stand-alone systems have proven to be a reliable, very low-maintenance method for producing electricity in moderate amounts in isolated, unmanned sites. These applications include remote communication repeaters and receivers; remote sensing stations; remote lighting systems; signals, including river and ocean navigational aids; and cathodic protection of remote bridges and pipelines. Such applications are too remote for regular maintenance or for fuel delivery. Experience with a variety of stand-alone systems has been excellent, with virtually no failures in any type of weather.

The hand-held calculator is an interesting example that illustrates the potential of portable stand-alone systems. One hundred and fifty million calculators worldwide are fully powered by light. The photovoltaic generator consists of nine solar cells connected in series. Each cell is about ¼ inch by ½ inch, and the entire array is ½ inch by 2 inches. This array generates

Figure 4-3 Radio relay equipment with solar module. (Courtesy: U.S. Dept. of Defense)

Figure 4-4 Van-mounted solar arrays used in the Department of Defense's telephone central van experiment. (Courtesy: U.S. Dept. of Defense)

Figure 4-5 Solar-lighted caboose. What's the cheapest way to light a caboose at night? With sunlight, of course. A square yard of photovoltaic array and a battery does the job for Southern Railway. (Courtesy: U.S. Dept. of Energy, with Southern Railway)

a few thousandths of a Watt at 3–4 Volts. There is no storage in the system, nor is there an on-off switch. When the calculator cover is opened, the light hits the solar cell, generates electricity, and the calculator is ready for use. After calculation is complete, the cover is folded over the cells, the generation ceases, and the system is off. This light-powered calculator is available for less than $20 retail. No batteries are ever required.

In the class of stand-alone photovoltaic systems, there are many portable, low-power (less than 100 Watts), lightweight applications. The list of portable photovoltaic systems currently available includes photovoltaic-powered digital watches, clocks, radios, televisions, and other portable electronic devices. More rugged applications include portable systems for charging the batteries of fishing boats and sailboats, caboose lights for trains, and military portable communications. All of these applications are now commercially available and economically sound. These systems supply power to their loads or alternatively provide trickle-charging for batteries to assure that electrical equipment always has adequate stored reserve, even when the system is infrequently activated or visited.

Figure 4-6 Natural Bridges National Monument in southeastern Utah. The 100-kilowatt photovoltaic array is located near the visitors' center on a 1.3-acre array plot characterized by a gentle slope to the south and an irregular shape to accommodate rock outcroppings. The variegated appearance of the array is the result of incorporating solar cell modules from three manufacturers. Toward the upper left of the picture can be seen the residence and maintenance areas. Within the maintenance area are the generator and the photovoltaic storage and control buildings. (Courtesy: U.S. Dept. of Energy)

Reliability and readiness are especially important in military applications. There are many situations in which a reliable, portable source of electricity is required. Figures 4-3 and 4-4 show two very different portable applications. One is a lightweight photovoltaic panel that charges rechargeable batteries for a portable field telephone system. Hundreds of these systems are being tested throughout our military. The other shown is a truck-transported mobile communications system. Vehicle-mounted photovoltaic systems can generate as much as 1 peak kilowatt from deployable fold-out panels. Experience has shown them to be safe and dependable.

Another excellent example of portable remote photovoltaic power is shown in Figure 4-5. The recent requirement that a railway caboose have a light presented a difficult, costly decision. Southern Railway chose to try a system built around a photovoltaic generator, battery storage, and a direct-current light. The photovoltaic array is less than 1 meter square and generates less than 100 Watts of electricity in a bright noon-time sun. All of Southern Railway's local-service cabooses are being converted to this solar-powered lighting. At first thought, a nighttime load for a daytime

Figure 4-7 A special plant has been built by Solarex Corporation, Frederick, Maryland, to produce semicrystalline (SEMIX) material. The facility is unique because all the energy required for its operation is solar energy derived from photovoltaic modules on the south-facing sloping roof. For this reason it has been dubbed a "solar breeder." It refutes the suggestion that more energy would be consumed to produce solar modules than could be obtained from them. In addition, it is a prototype for developing countries that do not have electric power at affordable costs to produce photovoltaic modules. The rooftop array produces 200 kilowatts in full sunlight. (Courtesy: Solarex Corporation.)

source could appear to be an odd choice. However, the photovoltaic system works well, has low maintenance, consumes no fuel, and is economic now for such small, portable electrical applications.

Another major set of applications for stand-alone photovoltaic systems are the fixed, large power systems. These systems presently include remote forest and Park Ranger sites, remote military bases, island locations, isolated residences, and remote villages. Today, these remote, large power systems are being built as experiments to prove that large systems can function with little maintenance, can be operated by semi-skilled persons, and can deliver electricity throughout the year with high reliability.

The largest U.S. stand-alone systems operating today are at the Natural Bridges National Monument Park in Utah (Fig. 4-6) and the Solarex plant in Frederick, Maryland (Fig. 4-7). The 100-kilowatt system at Natural Bridges is automatically backed up by a diesel generator and provides all the electricity for a small community of Park Rangers, park maintenance workers, and the visitor center. Almost fully automatic, it is the pioneer for many small community systems throughout the world. The basic design has been

Figure 4-8 Mt. Laguna facility. The 60-kilowatt photovoltaic power generation facility at Mt. Laguna, California, completed in June 1979, was made possible through a cooperative program between the Department of Defense and the Department of Energy.

The primary objective of the Mt. Laguna project is to demonstrate that a solar cell power system without energy storage can effectively augment a remote alternating current (AC) power network, providing reliable power with subsequent fuel savings. (Courtesy: U.S. Army)

improved in a 500-kilowatt system built using an advanced concentrator photovoltaic array in Saudi Arabia.

The 200-kilowatt Solarex "breeder" generates all the power for a new photovoltaic module manufacturing plant (Fig. 4-7). Dr. Joseph Lindmayer, the founder of Solarex, coined the term "breeder" because the plant generates more than 200 kilowatts of photovoltaic modules each year.

Another large photovoltaic unit is the 60-kilowatt system that augments a diesel-electric plant at a radar station operated by the Federal Aviation Administration at Mount Laguna, California (Fig. 4-8). This system could be a precursor for island, village, and community power systems throughout the world. The U.S. military is also using photovoltaic systems for refrigeration, signals, remote communication, remote target systems, and weapon systems power.

A photovoltaic-powered Indian Village project at Schuchuli, Arizona was perhaps the most exciting of Federal experiments to date. A small, 3.5-kilowatt

Figure 4-9 In the African village of Tangaye, Upper Volta, a 1.8-kilowatt photovoltaic power system grinds grain and pumps water less expensively than a diesel generator unit. Project sponsored by AID. Native labor was used to anchor arrays in trenches (left). (Courtesy: U.S. Dept. of Energy)

system (less power than is required for your home) powered water pumps, lights for 15 homes, a washing machine, lights for the community feast house, a sewing machine, and 15 four-cubic-foot refrigerators for 92 people in the village. The system was used for more than four years, until the Papago Electric Utility extended its service to the village in 1983. The Indian Health Service estimates that the United States has over 400 remote villages like Schuchuli and 80,000 individual Native American homes without electricity.

The U.S. Native American experiments are being copied throughout the world. The United Nations has estimated that there are two million villages worldwide like Schuchuli, i.e., they have no electricity and no electric utility available. The first international experiment modeled on Schuchuli was funded by the Agency for International Development (AID) in Tangaye, Africa (Fig. 4-9). In this application, the photovoltaic system provides fresh water and runs a grain mill to make flour for sale—a major change in the economy of this small community.

In developed and developing countries, one key economic factor in the agricultural application of photovoltaics is the provision of a constant supply to meet electric demands throughout the year. The University of Nebraska, the Kettering Foundation, Lincoln Laboratory of the Massachusetts Institute of Technology (MIT), and DOE have developed a photovoltaic-powered arc process for synthesizing ammonia and making fertilizer. The economics of small, remote applications, such as low-lift pumping, indicate that photovoltaics will soon be competitive with today's diesel generators.

Stand-alone remote applications, independent of an electrical grid, now involve relatively small installations far from central power. As costs decrease, photovoltaic systems can in time be built economically to provide virtually all of the electricity the United States requires.

Obviously, if electricity is desired when the sun is not shining, then we must store the solar electricity or have another generator on standby. The most common storage method today is sealed lead-acid batteries like those used on golf carts, industrial forklifts, and experimental electric cars. Battery storage works quite well with the direct-current photovoltaic array. The cost of storage, however, is another matter. It is about ten times break-even with power grid costs—lead-acid batteries now cost about $150 per kilowatt hour of storage capacity. They are also limited to about 800–900 refills (or deep-discharge cycles). In order to provide genuinely economic storage for photovoltaic systems, the battery industry must reduce costs by a factor of three and increase the discharge life to 2,000–3,000 cycles. Several new kinds of batteries that meet these goals are now under development by the government and industry.

Another storage option is to use photovoltaic electricity to spin a mass at very high speed. This flywheel storage can store enough energy in a 4-foot cube to provide a household with electricity for several days. New magnetic bearings and rotor materials are being developed to reach the cost/performance goals required for flywheels to be an effective storage method for use with photovoltaic systems. The storage industry has confidence that it can reduce the cost of storage so that reliable stand-alone solar photovoltaic systems can be designed and built to provide electricity 24 hours a day, day in and day out, for most areas of the United States.

Stand-alone applications, while important for what they can do, do not represent major energy-saving opportunities in the United States. Costs for stand-alone systems are too high. Systems must be designed and built that connect to and can supplement the trillion-dollar electric utility grid that carpets our nation. These "distributed" systems must be long-lasting, require little or no maintenance, have minimal impact on our environment, and be fully safe from the user's point of view.

GRID-CONNECTED DISTRIBUTED APPLICATIONS

Grid-connected photovoltaic applications are pivotal; they can save energy and fulfill many persons' desires to be somewhat independent of the utility while enjoying the security of having the grid as a backup.

The primary applications are roof-mounted arrays, facing south and tilted at an angle of latitude plus 10°. The roof array must have sufficient area to generate electricity to serve a large percentage of the building load. Fortunately for most one- and two-story nonindustrial buildings, the south-facing roof area receives enough solar energy so the array can generate electricity to serve a large share (greater than 60 percent) of the building's electrical requirements.

The most important on-the-grid applications, which can generate a large percent of the building load, include residences, mobile homes, two-story apartments, low-rise offices, small retail shops, schools, colleges, light industry, city and state buildings, barns, and covered parking. Few private experiments of this size have been built; entrepreneurs are waiting for photovoltaics costs

Figure 4-10 Mead, Nebraska, facility. An aerial view of the agricultural Flexible Test Facility, Mead, Nebraska. The photovoltaic array field is in the foreground. The experimental unit is operated by the University of Nebraska. The project was funded by the Department of Energy. The photovoltaic system was designed and constructed by MIT/Lincoln Laboratory. (Courtesy: U.S. Dept. of Energy)

to become more nearly competitive with the cost of power supplied by utility companies.

DOE, however, has funded a number of systems experiments to prove that multikilowatt grid-connected applications can be easily designed, installed, and operated with minimal maintenance. The first on-the-grid experiment was a 25-kilowatt irrigation experiment at Mead, Nebraska (Fig. 4-10), which was fielded by MIT's Lincoln Laboratory. The system has been operating in the extreme Nebraska heat, cold, rain, snow, hail, and wind, and it has provided reliable electricity for several years without degradation or extensive maintenance. In fact, the electrical grid that backs up the photovoltaic

Figure 4-11 The world's first photovoltaic-powered house in Phoenix, Arizona. The south-facing roof is composed of factory-sealed photovoltaic modules, which fit together with weather-tight connections. This DOE-funded project was built by Long Properties, Phoenix, using a system designed by ARCO Solar, Inc. The system generates 6–8 kilowatts when the solar flux is 1 kilowatt per square meter. This is enough electricity to cool the house to 78°F when it's 110°F outside, to light a few lights, to cover other normal daytime loads, and still have electricity to sell back to the local utility.

On dedication day, the photovoltaic system actually ran the meter backward while keeping the house cool with outside temperatures at 105°F. This particular array design provides the roof for the house as well as the electricity. The total labor to install the first system was only 40 hours. Ultimately, well-designed units will be installed in a tenth of that time. (Courtesy: U.S. Dept. of Energy) system has been blacked out three times more often than the photovoltaic experiment. The Nebraska system has also shown that photovoltaic power can be converted to alternating current and placed on the power lines without affecting the power grid adversely.

Three residential experiment stations (RES)—in Massachusetts, New Mexico, and Florida—provide controlled testing of systems for residential use. Photovoltaic arrays from different manufacturers, mounted on simulated housetops characteristic of the region, are being operated for extended periods of time while the power output is carefully monitored. Simultaneously, power usage is monitored at several nearby lived-in residences, and computers compare patterns of demand and supply. In addition, several makes of power-conditioning and -control equipment are being evaluated. The three RES facilities are providing a highly useful learning experience for all the interested parties.

Eight other test experiments have been constructed to evaluate intermediate-size systems for industrial-commercial applications. Each of these operates in conjunction with the local utility, which was involved in planning the installation, and which receives data as well as excess power from the unit.

How Much Electricity Does Your House Use?

Every utility bill indicates the number of kilowatt hours (kWh) used in the billing period. If a house does not use electricity for heating or air conditioning, it may use 350–600 kilowatt hours each month, depending on the number of appliances, the number of residents, and the patterns of electric power usage. For space heating and cooling, power usage could be more than twice this amount.

Let us assume that neither heating nor air conditioning is involved and that in a typical month your house uses 420 kilowatt hours, or 14 kilowatt hours a day.

A median insolation (sunshine) value for the United States is about 1 kilowatt per square meter at noon on a clear day. Considering cloud cover and haze, the total for an average day in, say, Virginia, Missouri, or Northern California is about 5½ kilowatt hours per square meter. If all this light could be converted into electricity, 3 square meters of array would provide the power you need.

Because the array is only 15 percent efficient in converting sunlight into electricity, the collector area must be nearly seven times larger, or about 20 square meters in area (about 215 square feet). Most single-family residences could readily accommodate an array of this size or larger on a suitably sloped portion of roof. This means that from the standpoint of size, photovoltaic systems are feasible for the single family at present-day photovoltaic efficiences.

Photovoltaics for New Housing

In the case of new housing, adequate roof area usually presents no problem, for it is relatively easy to assure a large, clear, south-facing roof of proper slope. Indeed, even most multifamily housing can provide this much roof space per dwelling unit. In some cases, two or three times this area could be allocated. This collector area would produce sufficient electricity to operate air conditioning and heating systems, especially if a heat pump is used.

From the point of view of energy efficiency, it is almost unforgivable to see expensive new housing being built that is not oriented to provide a large south-facing roof at a good angle to catch the sun's rays.[2] More often than not, otherwise satisfactory roof areas are chopped up with dormer windows or gables. The nation could ultimately realize enormous energy savings by one simple act: If every zoning jurisdiction in America promptly amended its ordinances to require adequate south-facing roofs clear of obstructions and with proper slope, new homes could exploit photovoltaics without major design changes or cost penalties.

Solar energy systems are commercially available that employ the thermal energy produced in the photovoltaic process, as well as the electricity, using photovoltaic concentrator designs. Compact dwelling units such as duplexes and multiunit apartments, where roof area is more limited, will find it advantageous to supply essentially all energy needs by using such total energy systems, even though they may be somewhat more expensive on a cost-per-square-foot basis.

Multistory apartments and office buildings, especially in northern regions, will make use of south-facing walls in addition to the roofs. Photovoltaic thin-film collectors will soon be inexpensive enough to make this a desirable strategy. In such cases, aesthetics becomes important. Photovoltaic collectors will be available as finished panels, in a variety of sizes and shapes, and with different kinds of trim, which can readily be incorporated either as wall or as roof in any new building.

As for existing housing, every case will be different, and ingenuity will be required in providing units for best economic return. A variety of units are on the market with mounts that permit arrays to be adjusted for existing varied roof and wall alignments, slopes, and spaces.

An analysis of the economics of photovoltaic use at a private residence is presented in Chapter Five and in Appendix 9.

The Cooperating Grid

A modest residential system with 250 square feet of array area, generating 3.5 peak kilowatts for new housing, may cost as little as $15,000 within a very few years. These grid-connected applications have led to serious analysis, technical and political, of the so-called utility interface issue. Can the electricity produced by photovoltaic systems be conditioned so that the quality of the electricity in the grid will not be affected? What is a fair price for the utility to charge when it backs up a photovoltaic user? When the photovoltaic system places excess power on the grid, what should the credit from the utility to the photovoltaic user be? Recent decisions by some state regulatory agencies indicate that the person owning a photovoltaic system will find it economically advantageous to have utility backup.

It is unlikely that our society would choose a noncentralized power system in which each home and business generated all of its own electricity. A diesel generator or an array of batteries would be needed at each site to supply backup power during the night hours and during overcast days. But this would be exorbitantly expensive. More economical would be a "cooperating grid backup" plan in which the utility companies would supply backup power. Such a plan implies a utility network whose pricing structure does not penalize the solar owner and residential photovoltaic systems that do not impair the functioning of the grid. Many system-integration studies by industry have shown that the photovoltaic phenomenon will not affect the grid dynamics, the grid reliability, or the grid reserve capacity until photovoltaic applications supply well above 15 percent of the total demand—and in some grids as high as 25 percent. If photovoltaics supplies more than 15 percent of the demand, the grid will have to obtain additional capacity through spinning-turbine reserve, or additional hydropower for storage, or storage mechanisms, including batteries. Storage is more economical if done at the grid load center rather than at individual houses and businesses.

Given a stategic approach to cooperating with the person who owns a photovoltaic system, the grid can effectively back up the solar option. If grid-connected photovoltaic systems obtain a very high penetration, greater than 15 percent, the utility may have additional costs. If society desires, the utility rate structure can be adjusted to distribute the additional costs to all customers. The value of renewable energy sources to the nation and the worth of solar as an indigenous resource far overshadow the added cost of grid backup. The cooperating grid scenario combines the best attributes of the small-is-beautiful philosophy with the best attributes of Amory Lovins's intense concern for appropriate use at the place of use, while making it possible for industry to continue to grow and make money.

The cooperating grid scenario is far from a flight of fancy. In California, Southern California Edison's rate demand charge means a person can actually run his meter backwards, use no electricity, and pay only $6 a month to have the backup of the grid assured at his home. The Public Utility Regulatory Policies Act (PURPA) of 1978 requires that a central utility pay for excess power supplied to it by a renewable source such as photovoltaics owned by a customer. The utility must also provide backup power for the solar generator. The exact rates for buying and selling power have been determined by the state utility commissions. PURPA specifies that the buy-back rates reflect the "avoided costs" to the utility for not having to generate and transmit the electricity.

The time has come for utilities to pioneer the cooperating grid scenario (as several are now doing), permitting the average person to own an electrical- and heat-generation system, operate and maintain it, be freed from reliance on the utility company for most of one's energy needs, and at the same time have the reliability of utility grid backup.

If there is a grid failure, another attribute of the cooperating grid scenario is that the homeowner still has some solar reserve when the sun shines. Given a daytime brownout, one could still keep the refrigerator, freezer, and other critical loads running. The addition of a few hours' worth of storage would provide emergency service in event of a power outage, occasional brownouts, or cloudy days.

CENTRAL UTILITY APPLICATIONS FOR PHOTOVOLTAIC SYSTEMS

There can now be little question that photovoltaics will be used directly by electric utilities. As approval for sites for large nuclear and fossil-fuel power stations becomes more difficult to obtain and pollution problems more expensive to resolve, and as the cost of photovoltaic systems comes down, utilities may be expected to turn increasingly to solar photovoltaics to satisfy peak power demands, which occur predominately during daylight hours.

Two unique projects are under way. The Sacramento Municipal Utility District has constructed a 1-megawatt unit adjacent to its Rancho Seco nuclear power plant. This first element of a proposed 100-megawatt photo-

Figure 4-12 This 1-megawatt central station photovoltaic system provides power to Southern California Edison at their Lugo Power Station near Hesperia, California. The supplier is ARCO Solar. A prototype for larger systems, this unit was constructed under "third-party financing." (Courtesy: ARCO Solar)

voltaic station was operational in mid-1984. The U.S. Department of Energy is providing one-third of the funds. The array is a flat-plate, fixed-tilt type. The second phase is under contract and will be completed soon.

In the meantime, the private sector has produced a unit entirely on its own. ARCO Solar, the world's largest manufacturer of photovoltaic modules, provided a 1-megawatt plant that supplies power directly to Southern California Edison (Fig. 4-12). The plant is owned by Atlantic Richfield Company, ARCO Solar's parent organization. The photovoltaic array consists of flat-plate modules mounted on pedestals that track the sun, which is somewhat more efficient in that climate. The entire system was constructed and placed in operation in less than a year. (ARCO Solar also completed a 6-megawatt central plant in 1984 at San Luis Obispo, California.)

The business arrangement is designed to attract investors who, as part owners under this third-party financing concept, realize some tax advantages in the short term and potential profits in the long term. Because the basic factor limiting the expansion of photovoltaics is the availability of capital, this somewhat unexpected recent third-party ownership trend should be healthy for the movement by increasing production and bringing prices down.

In similar arrangements, United Energy Corporation sold two large (greater than 1 megawatt) systems to investors in Davis and in Borrego Springs, California. In 1983, the company installed 5 megawatts of 2.5-kilowatt concentrator systems. The company plans to expand operations each year to annual sales of 50 megawatts. There appear to be no technical problems associated with multimegawatt central photovoltaic substations. Moreover, there is adequate land around coal-fired and nuclear reactor utility generating stations, and under transmission rights-of-way, to locate solar photovoltaic systems that could supply a large percentage of central electric power. The Electric Power Research Institute, General Electric, and Westinghouse have extensively studied the technical requirements, the interaction with existing generators, and the economics of this possibility. When system costs of $2,500 per installed kilowatt are realized, photovoltaic central power will be fully economic for the Sunbelt in this country. This is expected to occur in the early 1990s.

It is only appropriate that a review of photovoltaic energy systems conclude with a brief description of a concept that by any measure, be it distance, timing, magnitude, or sheer imaginative genius, must be rated the "farthest out"—the satellite power system, or SPS. Huge arrays of photovoltaic cells would be assembled into a large orbiting space platform. There they would collect sunlight and convert it into electricity, unimpeded by weather, the day-night cycle of the earth, or seasonal variations of insolation at the earth's surface (Fig. 4-13). As it was produced, the electricity would be converted into microwave energy, which would be beamed to earth from the satellite in geosynchronous orbit at a distance of 22,300 miles (36,000 kilometers).

The array for a single satellite would be some 25 square miles (65 square kilometers) in area, possibly larger, which could deliver to the ground perhaps 10,000 megawatts (10 gigawatts) of electrical energy. The receiving area would occupy about 26 square miles on the ground. People would not live in the receiving area, but a person would not be harmed walking through it. Approximately 25 such satellites could supply all the electricity the United States uses today.

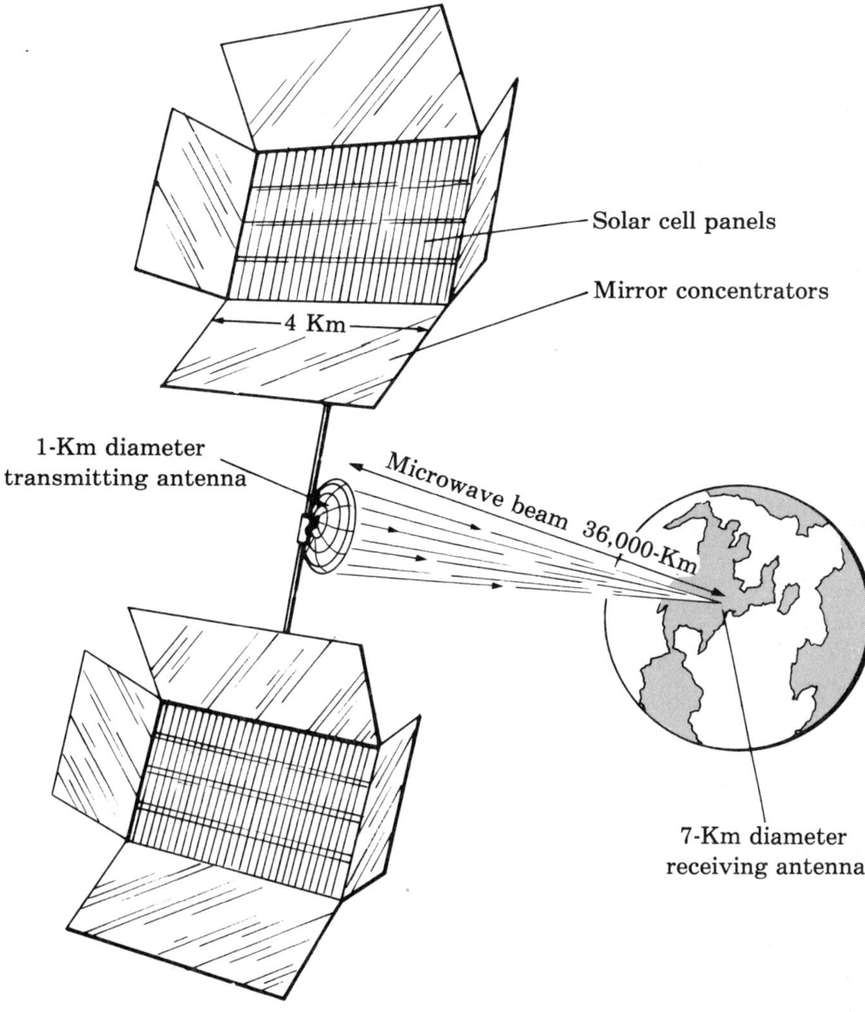

Figure 4–13 Solar power satellite concept.

Studies by DOE and NASA indicate that the concept is feasible. The concept has enough merit that a number of these satellites may in time be built to augment earth-based electricity sources. However, the technical and economic problems that might be encountered may well relegate the SPS to the next century. Space stations would be extremely vulnerable to military action, and a relatively peaceful world would be a prerequisite for investing heavily in such systems and then becoming dependent on them. By contrast, highly dispersed photovoltaic systems on the ground, giving each house a source of electricity at least for minimal needs, would be far more secure than the present utility network with its large central generators.

COMMERCIALIZATION BEGINS

Starting now, photovoltaic systems will find increasingly broad applications in the United States. Recognizing that new applications depend in part on the economic factors affecting the particular site and the owner, the general order in which photovoltaics may appear is suggested by the following list (most apt candidates first):

1. Remote villages and houses, existing and new
2. Third-party-financed ventures serving Sunbelt utilities
3. Small municipally owned utility systems
4. Remote water pumps for irrigation
5. Sunbelt residences with utility service
6. Residences in more northerly states
7. Educational institutions
8. Residential condominiums, high-density housing
9. Municipal buildings, hospitals, and other publicly and quasi-publicly owned structures

These early markets will enjoy explosive growth leading to a self-sufficient, billion-dollar industry by 1990. As costs drop, the late 1980s will see increasing economic viability for photovoltaics in light industrial complexes and utility installations near major load centers.

Because photovoltaic systems produce direct current, applications that can use DC power are favored candidates, especially if battery storage systems are already in place or if an intermittent power supply is acceptable. Such applications include water pumping for irrigation (no batteries); remote cooling, in which the energy produced is stored in the ice; water desalination; industrial processes such as the manufacture of fertilizers and some chemicals (no batteries); and highway lighting (with batteries).

As photovoltaic technology permeates society, the distinctions between particular applications will become increasingly blurred. Continued expansion will become more a function of proven economic benefits, of the reliability and convenience of photovoltaic systems, and of personal preference for the greater feeling of independence that comes from owning one's power supply.

DOE supported an extensive series of test applications of photovoltaic systems of various sizes, for a variety of uses, and under widely differing regional conditions, both as a means of perfecting the technology and in order to give the public an opportunity to observe photovoltaics in use.

Cost is, of course, the first consideration when most people decide whether to buy a photovoltaic system. But when such systems become generally affordable, other factors will come into play. A field assessment of factors that motivated homeowners in one area to procure solar hot water heaters, while not directly applicable, is indicative of factors that may affect photovoltaics:

- Existence of a local ordinance exempting the owner from property tax on the solar equipment
- Dwelling characteristics, particularly a favorable roof configuration (southern exposure and no shading)
- Perception of other private benefits, such as reduced fuel bills for a number of years
- Membership in environmental organizations and a desire to help with the energy shortage
- Encouragement from solar manufacturers, solar dealers, friends, neighbors, and household members

The study concluded that "certain incentives will trigger household decisions to acquire solar collectors when an appropriate social network encourages that commitment and when respondents are persuaded that private, not social, benefits accrue to the household from acquiring solar."[3]

NOTES

1. Photovoltaics history prior to the space age is reviewed in Appendix 1.
2. South of an imaginary line across the United States at roughly the Washington, D.C.-St. Louis-Denver-San Francisco latitude, a 45° southward roof slope is about optimum. North of this latitude, the slope southward should be 55° or 60° from the horizontal. The general rule is to add 10° to local latitude to obtain optimum year-round slope southward for a fixed array. An existing roof of any slope can be used if it has a generally southward orientation. Optimum slope becomes less important as the price of photovoltaic arrays continues to decline.

3. The factors that will motivate society to apply photovoltaic technology or any other solar technology are not well understood as yet—especially as related to the early stages of commercialization. This study, which examined the psychology of early market activity, may be of interest: Seymour Warkov, *Solar Adopters and Near Adopters: A Study of the HUD Solar Hot Water Grant Program* (Storrs: University of Connecticut, 1979).

FURTHER READING

Baldwin, Thomas E.; Hill, Lawrence G.; Santini, Danilo J.; and Stenehjem, Erik J. *Economic and Demographic Issues Related to Deployment of the Satellite Power System (SPS)*. PRC Energy Analysis Company and Argonne National Laboratory. ANL/EES-TM-23. Springfield, Va.: National Technical Information Service, 1978.

Beatty, J. Kelly. "Solar Satellite: The Trillion Dollar Question." *Science* 80, December 1980, pp. 28–33.

Bycer, Max. *Automated Solar Module Assembly Line: Quarterly Technical Report No. 5.* DOE/JPL 955287-5. Horsham, Pa.: Kulicke and Soffa Industries, Inc., 1980.

D'Alessandro, Bill. "Villagers Light the Way: Solar Cell Power in Gunsight, Arizona." *Solar Age,* May 1979, pp. 34–37.

DeMeis, R. "Solar Drones Scan the Earth." *High Technology,* January 1983, pp. 10–13.

Halacy, D. S. *Home Energy.* Emmaus, Pa.: Rodale Press, 1984.

Herwig, Lloyd O. "A Perspective of Solar Energy and Its Applications." *CRC Critical Reviews in Environmental Control,* July 1981, pp. 301–402.

Komp, Richard J. *Practical Photovoltaics.* Ann Arbor, Mich.: AATEC, 1981.

Kortman, P. J. "Flat Plate Photovoltaic Central Station Plant, Year 2000 Characterization." McLean, Va.: TRW, Inc., 1979.

Maycock, P. D. and Stirewalt, E. N. "Solar-Cell Systems That Work." *IEEE Spectrum,* September 1981, pp. 40–43.

Merrigan, Joseph A. *Sunlight to Electricity.* Cambridge, Mass.: MIT Press, 1975.

Metz, William D. "Energy Storage and Solar Power: An Exaggerated Problem." *Science* 200, June 1978, pp. 1471–1473.

Munjal, P. K., and Leonard, S. L. "Third Party Financing of Photovoltaic Power Plants for Electric Utility Service." El Segundo, Calif.: Aerospace Corp., January 1982.

Murray, William J. "Photovoltaics: Coming of Age?" *Solar Engineering,* January 1980, pp. 23–25.

"Ohio Daytimer Looks to the Sun for Its Power." *Broadcasting,* July 9, 1979, pp. 51–52.

Palz, Wolfgang. *The Prospects and Problems of Production, Storage and Terrestrial Applications of Electric Power Through Photovoltaic Solar Cells.* Prepared for the United Nations. Brussels: Commission of the European Communities, Directorate General for Research, Science and Education, 1979.

Photovoltaic Energy Systems: Program Summary. U.S. Department of Energy, Assistant Secretary for Conservation and Solar Energy. DOE/CS-0146. Washington, D.C.: U.S. Department of Energy, 1980.

Rappaport, Paul. "The History and State-of-the-Art of the Solar Cell." Princeton, N.J.: RCA Laboratories (undated). Available in mimeographed form from Photovoltaic Energy Systems Division, U.S. Department of Energy, Washington, D.C.

Rauschenbach, Hans S. *Solar Cell Array Design Handbook.* New York: Van Nostrand, 1980.

Riley, Julia D.; Odland, Robert; and Barker, Helen. *Standards, Building Codes, and Certification Programs for Solar Technology Applications.* SERI/TR-53-095. Golden, Colo.: Solar Energy Research Institute, 1979.

Ring, Randy. "Electricity from the Sun." *Country Journal,* December 1982, pp. 72–79.

Sanders, J. A. "The Concentrator Option." *Photovoltaics International,* June/July 1984, pp. 5–9.

Smits, Friedolf M. "History of Solar Cells." *IEEE Transactions on Electron Devices,* ED-23, July 1976, pp. 640–643.

Solar Electricity: Making the Sun Work for You. Technical Staff of Monegon, Ltd. Gaithersburg, Md.: Monegon, Ltd., 1981.

Solar, Geothermal, Electric and Storage Systems Program Summary Document: FY 1980. U.S. Department of Energy, Assistant Secretary of Energy Technology. DOE/ET-0102. Washington, D.C.: Government Printing Office, 1979.

"Solar Power Satellites." Office of Technology Assessment, Washington, D.C., August 1981.

Stewart, John W. *How to Make Your Own Solar Electricity.* Blue Ridge Summit, Pa.: TAB Books, 1979.

System Tests and Applications: Photovoltaic Program. U.S. Department of Energy/PRC Energy Analysis Company. HCP/T4024-01/15. Springfield, Va.: National Technical Information Service, 1979.

The Final Proceedings of the Solar Power Satellite Program Review, April 22–25, 1980, Lincoln, Nebraska. DOE/NASA Conf. 800491. Springfield, Va.: National Technical Information Service, July 1980.

"The Future of Solar Electricity, 1980–2000; Developments in Photovoltaics." Report No. M101. Gaithersburg, Md.: Monegon, Ltd., 1980.

The Solarex Guide to Solar Electricity. Technical Staff of Solarex Corp. (E. Roberton, ed.) Rockville, Md.: Solarex Corp., 1979.

Tucker, J. B. "Solar Power Goes On Line." *High Technology,* August 1984, pp. 44–51.

Turner, Rufus P. *Solar Cells and Photocells.* Indianapolis, Ind.: Sams & Co., 1980.

Warkov, Seymour. *Solar Adopters and Near-Adopters: A Study of the HUD Solar Hot Water Grant Program.* A Report to the Northeast Solar Energy Center, Cambridge, Mass. Storrs, Conn.: University of Connecticut, 1979.

Watts, R. L.; Smith, S. A.; and Mazzuchi, R. P. *Photovoltaic Product Directory & Buyers Guide.* New York: Van Nostrand, 1982.

Wolf, Martin. "Historic Development of Photovoltaic Power Generation." Philadelphia, Pa.: University of Pennsylvania (undated). Mimeographed copies available from Photovoltaic Energy Systems Division, U.S. Department of Energy, Washington, D.C.

CHAPTER 5
ECONOMIC FACTORS

The economic characteristics of photovoltaic use can be considered from the perspectives of the individual user, the photovoltaic manufacturer, the utilities manager, and the other players in the worldwide mixed economy in which solar cells are just one of many competing responses to the current energy situation. The changeover to photovoltaics and other renewable energy sources will accelerate as costs decline and as the nation's physical plant is steadily replaced. Photovoltaics will function as a technological "substitution" on the supply side of a domestic and world economy, and the magnitude of such a substitution is great: analogies are, say, the steamship for the square rigger or the car for the horse.

The substitution of photovoltaics for conventional electrical sources will proceed over time to the limit of usefulness, when we can expect another dynamic equilibrium to be established. Subsequently, photovoltaic use is likely to fluctuate marginally in response to microeconomic considerations.

How rapidly the substitution takes place and at what level market equilibrium will occur depend on two key factors—the federal government's policy commitments to alternative energy sources and public interest in photovoltaics.

There is, in essence, no technological limit on the use of photovoltaics.

There is, in essence, one physical limit on photovoltaic use: the amount of sunlight incident on each photovoltaic array.

There is, in essence, one economic determinant of photovoltaic use: the cost of electricity produced by other means.

There have been many forecasts for the economics of photovoltaics. Generally they conclude that by 1990, photovoltaic systems will be able to generate electricity that is competitive with conventionally fueled power in most parts of the United States.

One of the most succinct forecasts for the economics of photovoltaics is contained in a multiyear program plan issued by the U.S. Department of Energy in 1980 (Fig. 5-1).[1] It counted on continued federal support for

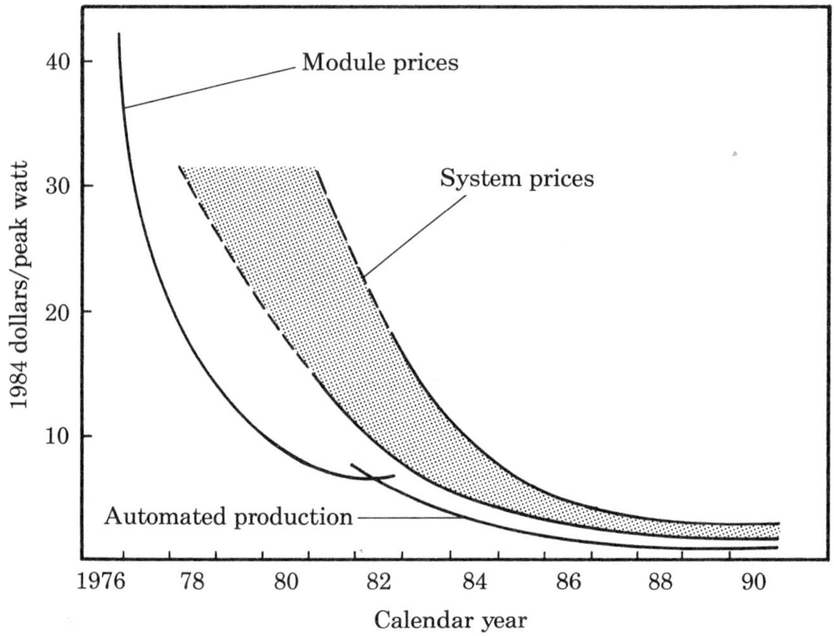

Figure 5-1 Photovoltaic module and system price goals made by the Department of Energy in 1980 (translated into 1984 dollars). These projections slipped significantly when oil prices declined and government support for photovoltaics was reduced.

research, development, testing, demonstrations, and market studies to assist the private sector in moving smoothly toward commercialization of the technology. Unfortunately, when oil became more plentiful in the 1980s and the energy crisis eased temporarily, the government relaxed its emphasis on all forms of renewable energy. The federal budget for photovoltaics was cut; as a result, the original DOE goals slipped by several years. When the government decided to limit its support to long-range research and development, the Department of Energy dropped its price goals altogether.

Nevertheless, if photovoltaic systems continue to drop in price as projected, they may provide several hundred megawatts of generating capacity within the next few years. By the end of the century, 5–10 percent of U.S. electrical demand conceivably could be supplied by photovoltaics.

Massive deployment of photovoltaics is inevitable. As the cost of producing solar cells declines and as the nation's physical plant is replaced, more and more users will choose photovoltaics on economic grounds alone. Forecasts differ on the question of how rapidly this deployment will take place.

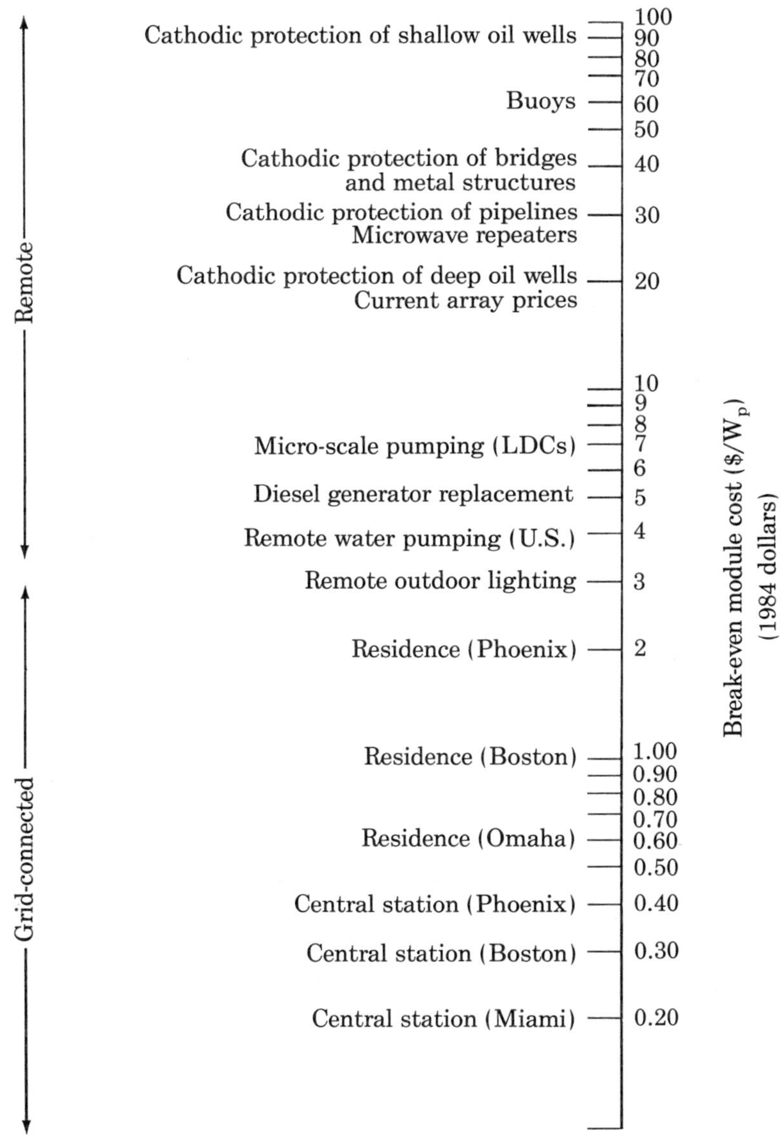

Figure 5-2 Photovoltaic applications and break-even costs. For each application shown, the amount indicated on the scale is the approximate price in 1984 dollars of photovoltaic modules that would make the electricity produced directly cost-competitive with electric power now used in that application. Equivalent system prices would be about double the module price. At the end of 1984, modules were selling for about $7 per peak Watt, systems for about $12 per peak Watt. (Modified from a chart by the U.S. Dept. of Energy)

The market for photovoltaic devices will develop in segments rather than all at once. There is no single price at which photovoltaics becomes economic. Photovoltaics is already competitive for some uses. As photovoltaic prices drop and the costs of alternatives rise, class after class of electricity users will turn to solar devices (Fig. 5-2). This sequential, segmental market pattern is typical of newly introduced technologies.

The actions of the government will be the most crucial in determining how rapidly the solar transition takes place. The government's role in the political economy is unique: By virtue of its power to tax, to spend, to mandate, and to prohibit, the government is a powerful force in determining the mix of energy sources the public will use.

Renewable energy sources provide our only real opportunity for restructuring the energy market on a scale large enough to lessen our dependence on all fossil fuels and especially on imported oil.

THE INTEREST GROUPS

Before looking at the particular situations of users and suppliers, and at the overall markets, a simple summary of the economic motives of all the parties involved may help explain likely market developments.

Users. Almost everyone uses electricity and uses it for a variety of purposes. But for whatever purpose, all classes of users are interested in acquiring needed quantities at low costs for operation, for maintenance, and for capital equipment. All users have a certain level of use below which their other activities are unacceptably hindered. Within a higher range of use, conservation measures are not burdensome, and there is a still higher level for each user beyond which consumption of electricity is clearly profligate. Users will want to get as much electricity as they can conveniently use and at as low a cost as possible.

An especially important way to divide users into subclasses is by their ability to afford the relatively large initial capital outlays that photovoltaic systems require. Many small businesses and lower- and middle-income households, in the absence of incentives, simply won't be able to allocate current income to install a photovoltaic array, even if they could realize significant savings over the life of the system. These users will benefit as the price of photovoltaic systems declines. For this to occur, others who are financially able to do so must support early market growth.

Photovoltaic manufacturers. The solar industry will want to see photovoltaic devices replace other energy sources, particularly the coal, oil, gas, and nuclear fuel used for electricity and heating. For the substitution to occur, photovoltaic manufacturers need to price their product competitively with other sources. Initially, they need to achieve great savings in the manufacturing process. The industry is now installing automated assembly lines.

Photovoltaic distributors. Distribution and installation of photovoltaic devices will likely be performed by small businesses whose economic motives will be conventional. Many existing heating, plumbing, and home improvement contractors may voluntarily seek the specialized, but relatively unsophisticated, training necessary to learn how to install photovoltaic systems in new and existing structures. A number of entrepreneurs specializing in solar systems will enter the field. A few nationwide firms, either new or existing, will enter the fray, using mass advertising and, eventually, their good reputations to attract sales; some will operate as franchises. Current energy suppliers, such as fuel oil distributors, may add solar sales and service capabilities in order to survive.

Local governments. States, counties, and municipalities typically wish to preserve and enhance the economic health of their areas, mainly by increasing the total spendable income of their constituencies. If solar systems can provide energy at lower cost than other alternatives, we can expect to see a series of tax incentives for solar installations designed to create jobs and to free up personal income for other uses. The solar industry could provide a number of new jobs, an important aspect of economic health at the community level. State and local measures to attract producers and to encourage distribution will be enacted. But states that are large producers of coal, oil, and gas could foresee economic loss if the need for these industries is lessened. Consequently, there may be at least a minimal antipathy at the state level.

Utilities and utility regulators. The posture of the utilities is the major unknown factor in the entire photovoltaic equation. Most, if not all, of the utilities in this country are monopolies regulated by public commissions, and they serve two functions, which were heretofore inseparable: the production of electricity and the distribution of electric power. We see distribution as remaining a natural monopoly. Our society will still need an electric-power grid, and it is wasteful to have more than one grid serving any specific geographical area. But the function of production could be radically altered: every photovoltaic user could be a generator. Whether the utilities will fight to preserve their function as the central and sole producer of electricity is an open question. Central generators will continue to be needed as backups during long periods of adverse weather and at times of extra-large peak demands. Utilities themselves can be expected to install centralized photovoltaic arrays for extra generating capacity whenever the cost of photovoltaic devices approaches the cost of other kinds of systems. Already, utilities are buying power from megawatt-size photovoltaic systems owned by third parties.

Regulatory agencies are generally charged with ensuring that a utility not earn more than a fair rate of return on its operations. Whether, during a massive change in the technological basis of the industry, many or most of these agencies will interpret this change as a mandate to protect the prior investment of utility owners is unknown but certain to be a public issue.

The federal government. The government's role, as stated, is unique.[2] We assume that government is responsive but not monolithic: Different parts react in different ways to different pressures, of which the economic interest of the single voter is but one. A chief motive of the government's formulation of an energy policy is to erase the potential threat to the political and economic health of the country from having large parts of its energy supply lie under the control of other sovereignties. Any strategy that lessens this reliance should be welcomed. Secondarily, the government appears to want to ensure growth of energy supplies (preferably at the least cost in development and operation), to create new jobs while not eliminating existing ones, to prevent environmental deterioration, and to enhance the physical health and safety of the public.

ECONOMICS OF PHOTOVOLTAICS FOR THE USER

The affordability of photovoltaics, now and in projections of future development, depends on how rapidly the cost of competing fuels rises; the speed with which photovoltaic hardware is developed; and the strategies the federal government uses to foster, or to discourage, the inevitable changeover to photovoltaics. Projections depend on the definition of costs, the values assigned to external costs such as pollution or storage of radioactive waste, and such intangibles as aesthetic considerations and convenience. The effective price depends on the money market and the user's tax status—the numbers look very different from the respective viewpoints of the commercial utility, the private homeowner, the renter, the landlord, the retail business owner, the manufacturer, and the non-profit institution.

Despite the complexity of the situation, two factors are paramount. For all users, the economic attractiveness of photovoltaics will depend primarily on the combined effects of insolation rates and local electricity prices. Insolation (the amount of available sunlight in a particular location) determines the size of an array needed to produce a desired amount of current; thus it is a main determinant of the installed capital cost of a photovoltaic system. A house in New England would require a system about 40 percent larger than the same house in Arizona, excluding the climate-influenced differences in the power required for heating and air conditioning. See the appendix titled "Estimating Photovoltaic System Requirements for a Residence."

Local electricity prices are the key variable in operating costs. The rates at which utilities not only sell but also buy back electricity appear to be critical factors in determining photovoltaics' economic viability. The decision to go solar turns on the question of whether the user will save enough money by not buying electricity to offset the price of installing photovoltaic equipment.

Economic factors such as the tax posture of the purchaser, interest rates, and government incentives play a lesser, though significant, role in determining how economically attractive photovoltaics will be. Particularly during the next decade or two while the technology is new and the array price is high, these other factors will comprise the determining margin. The most likely residential purchaser will be the homeowner (but not the landlord) who lives in a region with abundant sunlight, is in a high tax bracket, can borrow money at low rates, and receives a large credit for the installed price against his federal income taxes. Others less favorably situated would delay a decision to convert.

Photovoltaic systems are characterized by relatively large purchase costs, low operating and maintenance costs, and free fuel. Thus, any user considering a photovoltaic system will have to compare the purchase cost, amortized over the life of the system, against the projected expenditures for buying electricity from a central utility. High purchase costs work against those who cannot afford to wait to recoup their investment and those, such as renters, who have no capital invested in the site and do not expect to stay long enough to realize any savings on operating costs.

A look at the numbers involved in two uses of photovoltaic power, replacing a diesel-fueled generator and supplying a U.S. residence, provides an indication of the price ranges at which photovoltaic systems become competitive.

The Small Portable Generator

Remote stand-alone or portable photovoltaic systems are now fully economic when used to recharge, or extend the life of, battery systems. About 3 megawatts of photovoltaic systems are sold each year for this use, and the world market could grow to as much as 500 megawatts per year.[3] At present, battery systems are frequently recharged by small diesel or gasoline generators. These generators ordinarily produce less than 50 kilowatts of

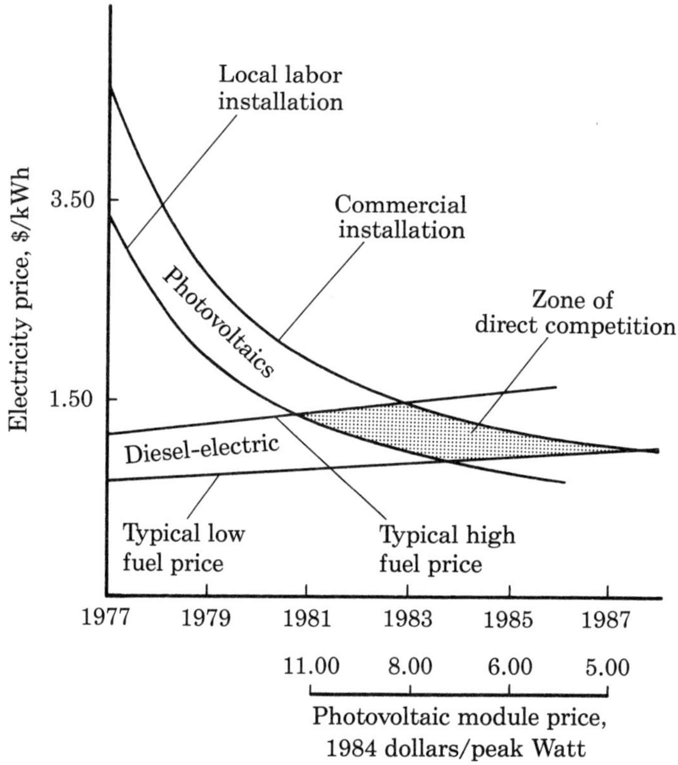

Figure 5-3 Energy price comparisons for diesel generator and photovoltaic 1-kilowatt continuous power systems. This figure compares the energy prices of a 3-kilowatt diesel generator and a 5.5-peak-kilowatt solar photovoltaic system with battery storage. Both produce 1-kilowatt continuous power, and the photovoltaic system requires relatively good insolation. The two diesel curves reflect fuel prices of $0.60 and $1.50 per gallon, rising 5 percent per year. The two photovoltaic curves reflect the use of local noncommercial and commercial labor, costing 10 percent and 50 percent, respectively, of the total installed system cost. Both photovoltaic curves reflect the module prices shown on the lower scale. The overlapping sections of the curves denote the economic break-even zone. Diesel fuel prices have been rising faster than the projections shown here, and photovoltaic systems are now producing electricity competitively with small diesel generators in many parts of the world. (Modified from a chart by NASA Lewis Research Ctr.)

electricity and are used for such purposes as powering small remote villages, small radar installations, and mobile Army units.

Break-even cost analyses for a diesel generator compared with a photovoltaic system have been developed in detail by the NASA Lewis Research Center[4] and by Royal Dutch Shell in the Netherlands.[5] Both studies concluded that photovoltaic systems are competitive with small diesel generators when the installed photovoltaic cost is less than $8 per peak Watt.

The principal cost of operating a diesel generator is the fuel it consumes. To produce 2,000 kilowatt hours of electricity, a well-maintained 3-kilowatt diesel generator (producing 1 kilowatt of continuous power) burns about 400 gallons of fuel.[6] Figure 5-3 shows the effect of fuel cost on electricity generated in this way.

The average price for delivered diesel fuel in remote parts of the world's Sunbelt is $3–$5 per gallon. This means that the 10 million small diesels in use throughout this area are generating electricity at a fuel cost of $0.60–$1.00 per kilowatt hour. If we amortize the cost of the generator and add in the cost of spare parts and regular maintenance, the cost of this electricity is higher, typically ranging from $0.75–$1.50 per kilowatt hour.

If the same amount of electricity were generated by a photovoltaic system, a complex new set of economic factors must be considered. These include the amount of sunshine available, the presumed lifetime of the photovoltaic system, the cost of maintenance, the cost of storage (if any), the cost of money, and of course, the unit cost of the system. For a simple system without storage, let us assume:

- A Sunbelt site where the annual solar flux is sufficient for the photovoltaic generator to produce 2,000 kilowatt hours per year per kilowatt of generating capacity
- A 20-year loan with a 15 percent interest rate
- A photovoltaic system life of 20 years
- Maintenance costs of 1 percent of capital per year
- Peak capacity of 1 kilowatt (photovoltaic systems have little or no economy of scale)

If the installed capital cost is $10,000 and this 1-kilowatt generator produces 2,000 kilowatt hours per year, to find the cost of electricity generated, the annual cost of capital plus maintenance is divided by the system's output in kilowatt hours per year.

60-kilowatt photovoltaic system
Lifetime: 20 years—131,400 kWh/yr
2,366 modules (97,000 cells)
½-acre array
20-year cost (1980 dollars):

Capital investment (land additional)	$600,000
Fuel	0
Total	$600,000

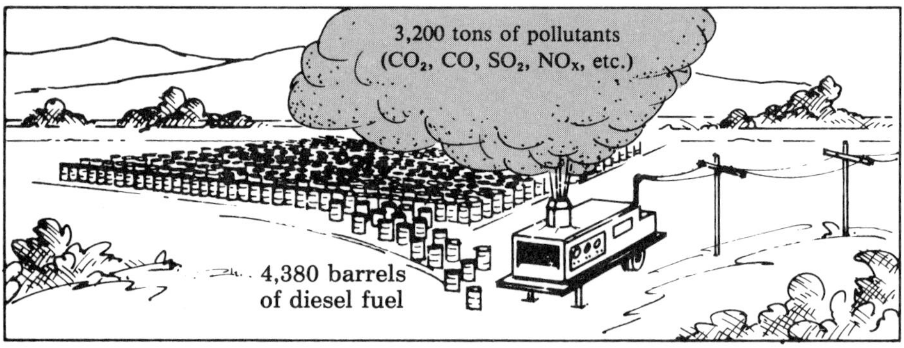

60-kilowatt diesel-electric generator
Lifetime: 20 years—131,400 kWh/yr
Diesel replacements: 10 @ $16,000 each
Fuel: 9,000 bbl (1,039 tons)
20-year cost (1980 dollars):

Capital investment	$160,000
Fuel (@ $3/gal)	$1,134,000
Total	$1,294,000

(operation and maintenance extra)

Figure 5-4 Comparison of photovoltaics with diesel-electric for power generation in remote areas. Because of the high price of diesel fuel in developing countries, photovoltaic systems are fully competitive with diesel-electric generation for many areas. (Idea courtesy of Andrew Krantz)

The annual capital cost to amortize the mortgage and maintain the system is nearly 20 percent of $10,000, or $2,000. Thus, the cost of electricity per kilowatt hour is $1. This compares favorably with the cost of electricity from diesel generation as indicated in Figure 5-3.

Large diesel generators are more efficient than small ones. At installed costs of $8 per Watt, photovoltaic systems would be competitive with small diesel generators (5 kilowatts or less). At $5 per Watt, photovoltaics will compete with diesels in the 20-kilowatt range, and at $3 per Watt, with the large 100-kilowatt diesel-electric sets.

The Private Residence

Photovoltaics is just beginning to be introduced for grid-connected uses, but it is now the preferred source of electricity in houses that have no utility service. Nearly a thousand residential photovoltaic systems will be purchased for private use this year, and by the end of the century, millions of residences will be served by individual, semi-independent photovoltaic systems. (To some extent, this discussion can be generalized to commercial and industrial applications as well, since the expense of the photovoltaic plant can be amortized and deducted as a legitimate expense of doing business.) How is this photovoltaic penetration into the market going to come about?

Photovoltaics will be competitive in many areas of the market when module costs approach $3 per Watt and installed systems are $5 per Watt. Because of rapidly changing fuel prices, housing costs, mortgage rates, and incomes, it is nearly impossible to describe a "typical" residential situation. Still, in order to understand how these variables interact in determining when photovoltaics becomes economically feasible, discussing a representative case can be useful. Assume that your house uses 24 kilowatt hours of electricity on an average day, or 8,760 kilowatt hours a year. Assume further that you want a photovoltaic system that will provide about 80 percent of your annual power consumption, or 7,000 kilowatt hours a year. You live in the Sunbelt, so a photovoltaic system can deliver 2,000 kilowatt hours a year for each kilowatt of installed capacity. Dividing 7,000 by 2,000 gives 3.5—thus, 3.5 kilowatts of capacity is required.

At an installed cost of $5 per peak Watt, the system would price out at $17,500. Assuming that state and federal tax credits combine to equal $5,000 ($4,000 federal credit plus $1,000 from the state), the cost to you would be $12,500 after taxes.

If the interest on this is 15 percent, and you are in a 35 percent tax bracket, the net interest rate to you would be 10 percent, or $1,250 a year. The cost per kilowatt hour would then be $1,250 divided by 7,000 kilowatt hours, or $0.18 per kilowatt hour. (Without tax credits, this would be $0.25 per kilowatt hour.)

Under such circumstances, it is obvious that not many homes with ready access to the grid will be equipped with photovoltaic systems soon; \$0.18–\$0.25 per kilowatt hour is simply too expensive. However, rising costs for electricity supplied by the utility companies may soon change things dramatically. Many experts predict a tripling of utility charges for electricity in the next few years. Thus, if you are now paying \$0.07 per kilowatt hour, your bills may rise to \$0.21 per kilowatt hour. Photovoltaic electricity will then be quite competitive. It's also important to realize that photovoltaics is the one energy source that will actually decline in cost. While energy costs in general will rise in the years ahead, the cost of photovoltaic electricity will actually decline, so photovoltaic systems may ultimately provide electricity at lower rates than the utility companies will charge.

Another factor to bear in mind is that the \$0.18–\$0.25 charge for photovoltaic electricity results from the high interest rate we assumed (15 percent). If interest rates were lower, the price of photovoltaic power would fall proportionately. For example, if we assumed an interest rate of 5 percent instead of 15 percent, the cost for photovoltaic power in our example would be just \$0.06–\$0.08 per kilowatt hour.

For homes that do not have ready access to the grid, photovoltaic systems are already the preferred energy source. If your home is even a few hundred yards from the nearest power line, the cost of extending the line to your house could be several thousand dollars—higher if the land is hilly, rocky, or heavily wooded. In such cases, installing a photovoltaic system often makes far better financial sense than hooking up to the grid.

In 1985, installed residential photovoltaic systems cost \$8,000–\$12,000 per kilowatt hour peak capacity. Forty percent federal tax credits for purchasers of photovoltaic systems make photovoltaics competitive in many more locations, provided the utility company cooperates by providing backup electricity at a fair price.

The cost of photovoltaic systems, the cost of money, the purchaser's tax status, and the cost of grid electricity are all parameters that must be factored into the decision as to when purchasing photovoltaics is economically advantageous. Table 5-1 compares key factors affecting five different groups of potential users in the Southwest. In one important sense, the homeowner is actually in a more favorable economic position to buy a photovoltaic system than other users of electricity or than electric utilities that produce it. He or she can borrow money at lower rates by deducting the interest paid on a loan in computing personal income tax, and in addition, when solar tax credits are available, he is allowed to subtract directly 40 percent of the cost of the photovoltaic system from his or her computed income tax. If tax credits for homeowners are extended, residential applications should eventually represent the major domestic market for photovoltaics.

The industrial customer or central utility must pay 2-3 percent more than prime rate for new capital. This is a relative near-term disincentive for the business or utility concerned. On the other hand, a university or other school can obtain construction funds for as little as 3 percent interest and spread repayment of the loan over 40 years.

From the point of view of a central utility manager, the value of electricity with which photovoltaic costs should be compared is his marginal cost of generating electricity from a new plant that uses some alternative source: coal, nuclear, hydro, whatever. That is, the comparison is in terms of the cost of generating more electricity from a new plant. In the last few years, estimating the cost of new central station plants has become a hazardous game for the utilities. Coal plants must have scrubbers. Nuclear plant costs have escalated to the point that several utilities have been forced to the brink of bankruptcy.

Another disincentive for the manager in business or industry is the fact that he can deduct all electricity costs from his income; in effect, he is receiving a 45 percent rebate from the IRS for using non-renewable fuel. If he were to purchase capital to create electricity from the sun, however, IRS may allow as little as 5 percent of capital per year to be depreciated.

Solar investment tax credits can mitigate this factor somewhat. Tax-exempt institutions, because of their status, can deduct neither power costs nor capital. The homeowner can deduct all interest on the home loan to purchase

Table 5-1 Key Economic Factors in the Purchase of Photovoltaic Systems Compared with Grid Electricity in Southwestern United States

Factor	*Residence*	*Industry*	*Central Utility*	*Municipal Utility*	*College or University*
Cost of grid electricity	5-9¢/kWh	2-5¢/kWh	3-4¢/kWh	3-4¢/kWh	6-8¢/kWh
Cost of money* (Prime less 3%)	9% (Tax rebate in addition)	14% (Prime + 2%)	15% (Prime + 3%)	3-5%	5-7%
Term of loan (yrs)	30	20	20	40	40
Land or roof cost	Free (no extra cost)	$1/sq ft	$1/sq ft	Free to some	$1/sq ft

*Prime rate of 12% is assumed for purpose of illustration.

the photovoltaic system, but cannot deduct the cost of purchasing electricity from the utility.

Obviously, identical photovoltaic systems can have very different costs and benefits, depending on the purchaser's status vis-à-vis a number of variables. There are a range of applications to which photovoltaics can be put, each of which must be analyzed separately. Within each class of applications, the tax structure, utility costs, and income of the owner all collaborate to determine the break-even point. The development of the industry itself needs to be examined within the context of a global mixed economy; there are more variables, this time on a macroeconomic level. Photovoltaic viability on a scale large enough to seriously relieve our dependence on non-renewable fuels is intimately tied to the size of its market.

The grid-connected residential market will develop slowly. It may be expected to boom only when interest rates come down and photovoltaic system prices decline significantly. The latter will result from technological advancements that increase efficiency and from manufacturing improvements that reduce production costs. Increasing use of photovoltaics for stand-alone residential systems will help the process along.

ECONOMICS OF PHOTOVOLTAICS ON THE SUPPLY SIDE

A market is made up of both buyers and sellers. Multiple economic considerations affect both manufacturers of photovoltaic equipment for applications on earth and utilities in their role as suppliers of electricity.

In the 1960s, when solar cells were used mostly on spacecraft, five companies entered the field: Heliotek, Hoffman (later Centrallab, now Applied Solar Energy), International Rectifier, RCA, and Texas Instruments. The annual market grew from 7,000 Watts in 1960 to a steady level of 80,000 Watts in 1968–1970, selling at between $100 and $200 per peak Watt.[7] Along the way, the latter three companies dropped out of photovoltaic production. A simple case of industrial overcapacity had developed: There were too many producers and not enough places, at current prices, to sell.

In the early 1970s, an awareness of a potential market on the earth spread through some segments of industry, the public, and the government. By 1979, the U.S. photovoltaic industry comprised more than 100 companies pursuing five main activities: raw materials production (typically upgrading raw silicon to semiconductor-grade crystalline form), photovoltaic components production (typically producing silicon wafers from crystalline ingots), photovoltaic systems manufacturing (assembling photovoltaic modules and arrays), original equipment manufacturing (installing electric power sources on such products as wristwatches and portable refrigerators), and research and development in all the above phases.

The process begins with raw silicon. Hemlock is the largest domestic producer of semiconductor-grade silicon. Its 900-metric-ton production in 1984 comprised about half of the output for the United States and a fourth of the world total. (Wacker in Germany produced 2,000 tons.) Other producers include Texas Instruments, Monsanto, Motorola, and Great Western. About 1 percent of this silicon is used by photovoltaic manufacturers, and the cost of the silicon accounts for at least half the cost (not price) of the finished cell.[8] This manufacturing process is energy- and capital-intensive, and new production capacity using the standard Siemens process requires a long lead time (typically three to five years). Prices of purified silicon are expected to fall from $60–$20 per kilogram by 1988.

About 20 U.S. firms supply photovoltaic components. Mainly, these companies produce silicon ingots and slice them into wafers. Some 15 U.S. firms manufacture silicon photovoltaic systems for use on the earth; two others limit their sales to space applications.

Most of these companies carry module production forward into the final assembly and sale of photovoltaic arrays; they also sell modules to other companies, which specialize in final assembly. Module prices in 1984 ranged from $5–$9 per peak Watt and installed systems from $8–$20 per peak Watt. World sales in 1984 totaled 30 megawatts worth $210 million. A breakdown of the 1983 world market and the position of U.S. firms in it are shown in Table 5-2.

Table 5-2 Position of U.S. Firms in World Photovoltaics Market in 1983

	Market Share (%)	*Shipments (kW)*
ARCO Solar	27.7	6,000
United Energy	20.8	4,500
Solarex	6.0	1,300
Solar Power	1.8	400
Solec International	1.4	300
Photowatt International	1.4	300
Mobil Solar	0.3	70
Others	0.9	200
Subtotal U.S.	60.3	13,070
Europe	15.2	3,300
Japan	23.1	5,000
Other	1.4	300
Total World	100.0	21,670

Another segment of the industry, original equipment manufacturers, is concerned primarily with the reliability and cost effectiveness of its products.

Some products now use photovoltaics, and more will do so when photovoltaics meets these criteria.

Finally, some 25 domestic photovoltaic companies are engaged solely in research and development activities. Several depend heavily on federal funds. Many have a weak commitment: If funding were ended, they would drop photovoltaics work. We identified $82 million of privately funded research in 1984, mostly oil-company funded.

Several unresolved energy issues make it difficult to forecast the characteristics of a mature photovoltaic industry. One of these concerns ownership and competition. Of the 15 or so silicon module manufacturers, several are independent, several are subsidiaries of electronics companies, and two are oil company subsidiaries. ARCO Solar, formed through acquisition by Atlantic Richfield, and Solarex, owned by Amoco, are actively selling photovoltaic arrays. Other oil companies, including Chevron and Gulf, have launched R&D efforts. Shell Oil participates in a joint venture with Motorola.

There is no question that Big Oil is becoming Big Energy. The energy industry today means oil, gas, coal, and nuclear. In 1984, it accounted for 20 percent of the gross national product. The oil giants now own and produce most of the coal and a significant fraction of gas and nuclear electric power, and they have also diversified into other energy and non-energy areas. In recent years, Big Oil, now Big Energy, has been quietly buying up small solar energy independents.

Expansion into photovoltaics is a logical step for the oil giants to take for several reasons:

1. Over the long run, oil supplies will dry up, but the need for energy will not.
2. Like oil exploration, photovoltaics is a long-term investment and presents no peculiar management problems.
3. Oil companies are among the largest in the country. Large profits force them to consider alternative investments and enable them to sustain money-losing ventures for a long time. Photovoltaics presents sound opportunities for investment in the energy field, which they have the capacity to underwrite.

Although small photovoltaic companies have formed in recent years, the trend toward concentration in the energy industry is clear. Is this trend healthy, from the standpoint of providing solar photovoltaic equipment to the marketplace at reasonable prices and in the quantities that the American people want? More to the point, will the new owners, who obviously have other interests, give solar a fair boost into the marketplace? At this stage, there is no way to be sure. Much will depend on government attitudes, but even more may depend on the public's professed desires.

However, the presence of both electronics subsidiaries and independents in the photovoltaic module industry indicates a diversity, which suggests that both technological and price competition will continue unabated over at least the short and medium run.

Furthermore, the cost of entering the field is low. At present, economies of scale in flat-plate array assembly level off at plant sizes of between 30 and 50 peak megawatt annual production. To start a plant of this size will require an investment of only \$10–\$15 million in the next few years.[9] Since entry cost is low, however, independent companies in the field are always vulnerable to takeover.

A second issue concerns the economic impacts of the rapidly changing technology in the field. Only two module production methods are approaching anything near maturity: single-crystal silicon and polycrystalline silicon. Whether these rapidly evolving technologies or some of the others now in earlier states of investigation will become dominant in the next few decades is not now known. Amorphous silicon plants are now operational. It will take two to four years before this option matures.

The rapid pace of change yields several uncertainties. Potential investors in a new enterprise may be unwilling to gamble on an operation whose technology may soon be obsolete. Also, no one knows what the most efficient scales of operation for both present and future technologies will be, and this scale is the single most important determinant of the future structure of the industry. One study speculated, "It is possible that silicon materials production could become most efficient at large scales, module production at medium scale, and residential systems sales, installation and service at a scale attractive to small business."[10]

In any event, the industry is young enough and the technology changing so rapidly that the economic advantages gained by small firms' proprietary research and development findings could offset the advantages held by large firms' capital resources.

A third issue concerns the nature of vertical integration of firms in the whole photovoltaic production process. Manufacturers of metallurgical silicon may move to produce crystalline silicon and vice versa. Crystalline silicon manufacturers may move forward into wafer, module, and system production. Original equipment manufacturers may move all the way back to module production. These moves will be made for financial profit and, in some cases, to ensure supplies of manufacturing components. Again, economies of scale will dictate industry structure, and it remains to be seen where these economies will lie.

A final issue concerns government policy. The government has not yet made some of the decisions that, by virtue of its role, it will have to make. These unmade decisions will also affect industry structure. The imminent economic viability of photovoltaics has not even become common knowledge among parts of the Department of Energy, much less among other departments involved in energy issues, or Congress generally.

DISTRIBUTION OF PHOTOVOLTAIC DEVICES

Distribution is the final step in getting photovoltaic devices out of the research journals and onto rooftops. Distributors have the task of locating potential purchasers and installing devices on unique sites. The problems are very similar to those faced by the heating and cooling industry, and we can expect photovoltaic distribution to resemble that model closely, if not to become an actual adjunct.

A mix of the following seems likely.

- Photovoltaic systems manufacturers may wholly own or may franchise a network of distributors.
- Independent distributors may stock a variety of brands.
- Existing heating, air conditioning, plumbing, electrical, and home-repair and -improvement firms may add photovoltaic capabilities.
- A number of photovoltaics specialists could appear.
- Large national retail firms may market photovoltaic systems and provide installation and maintenance services.

The construction industry could serve as a natural conduit for bringing photovoltaics into widespread use. However, the entire industry, including private contractors, financers, and public regulatory bodies, is notorious for its institutionalized lack of innovation. Even after the cost of photovoltaic electricity becomes fully competitive with electricity from the utilities, the construction industry can be expected to move slowly.

There will be barriers to photovoltaic deployment within the construction industry due to various factors:

- Housing builders are typically very small businesses, and have a strong tendency to keep doing only what they know how to do.
- There are 30,000 building code variations spread among the nation's 4,000 code-making authorities, and these codes tend to be oriented to specifi-

cation rather than performance: One must "do it this way with these materials" rather than "achieve this result."

- Planning and zoning boards and building code inspectors, one by one, will face demands for photovoltaics; many will resist.
- Bankers don't like to lend money for unknowns; news of proven photovoltaic performance will seep slowly through the financial community.

These barriers are institutional rather than technical or economic. Under pressure from consumers, manufacturers, and national energy needs, such obstacles will fall—slowly and grudgingly, perhaps, but they will fall.

UTILITIES

The utilities' physical and economic positions are unique in any energy analysis. The utilities act as both producers and distributors of electric power and serve all areas except where the cost of erecting new transmission lines is prohibitive (new users may bear this cost if they wish).

Because it would be a massive waste to permit the erection of more than one set of central generators and transmission lines to serve a single area, governments have granted utilities the right to operate as monopolies. To protect the economic interests of consumers, governments have also created public regulatory bodies with power to establish sales prices and the responsibility to supervise operations. Utilities are required to provide power to anyone in their area who wants to buy it, which explains their need always to have ample generating capacity. Similarly, utilities are eager to urge conservation if a power shortage is foreseen. Investor-owned utilities are allowed to make a reasonable return on their investment.

Figure 5-5 depicts the distribution of power generation by ownership. Production for the last decade is shown in Table 5-3. While vigorous conservation efforts, coupled with fast-rising electric power costs, have curbed the rate of growth since 1978 (from 7 percent a year to a little more than 3 percent in 1984), nevertheless steadily increasing usage is foreseen. For example, Besmann and Fox of the Oak Ridge National Laboratory point out that, as the public switched away from oil and natural gas, the number of homes heated primarily with electricity increased from 11.5 percent in 1973 to more than 17 percent in 1984, and more than half of all new houses now being built use electric heat.[11] By 1990, they say, electricity consumption in homes and businesses will be 25 percent greater than in 1983, and industry will require 35 percent more electric power. They warn that, by the early 1990s, the nation's electric power system will fall below the 20 percent reserve margin considered prudent by the utilities during periods of peak demand.

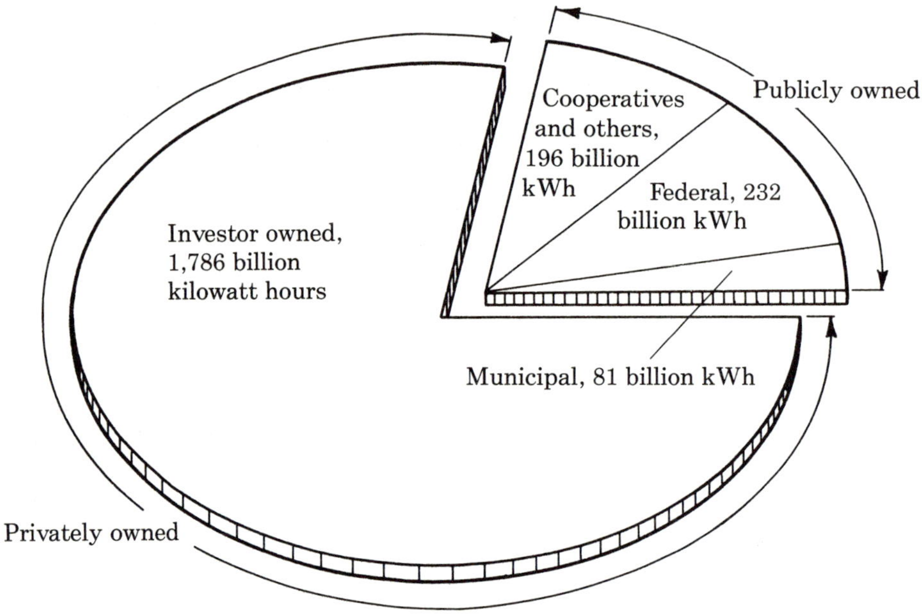

Figure 5-5 1981 utility-produced electricity for public use, by ownership. (Source: Statistical Abstract of U.S., 1982–1983)

Against this picture of growing use must be arrayed significant possibilities for further conservation through such foreseeable opportunities as doubling the efficiency of electric motors, tripling the efficiency of lighting, and quadrupling the efficiency of major appliances such as refrigerators, furnaces, and air conditioners. Energy consumed in heating and cooling of buildings may be reduced 10- to 100-fold, and big savings can be made in industry through such things as smelting processes that consume 40 percent less electricity and on-site cogeneration using otherwise wasted heat.[12]

In any event, it seems clear that without photovoltaics there will be heavy pressure to build additional large coal and nuclear plants. But progress with photovoltaics now virtually assures that adequate power will be available from clean, safe solar energy.

In countries that already have a well-developed electrical grid, photovoltaics will not alter either the need for, or the basic expense of, maintaining and operating the electric distribution network. We can assume that because of its convenience, the transmission grid will continue to be maintained, consumers will bear the costs, and public agencies will continue to regulate.

Extensive use of distributed photovoltaic systems will, however, alter the nature of electric power generation. As in the case of the potential residential user, many variables enter into the economics of central power generation. While precise analysis is difficult, the major factors are clear enough for us to say with confidence that photovoltaics could benefit both user and utility.

The two major costs of central power generation stem from plant capacity (the cost of constructing and maintaining generating facilities) and from fuel (the main cost of operating them). Photovoltaics will affect both but by differing degrees. Such conditions as time of day, season, latitude, cloud cover and available sunlight, and total amount of photovoltaic electricity production affect the economics of photovoltaics.

Utilities must meet three fairly distinct levels of power demand: base load, which is the minimum amount required across a 24-hour period; intermediate load, the additional amount required to cover normal daytime surges; and peak load, needed to meet extra-heavy demands imposed by heating loads on very cold days and by air conditioners in very hot weather.

Table 5-3 U.S. Utilities' Electricity Production for Public Use

Year	*Total Production (billions of kWh)*	*Change from Previous Year Listed (%)*	*Per Capita Production (kWh)*	*Change from Previous Year Listed (%)*
1973	1,860	—	8,900	—
1974	1,867	0	8,800	−1.1
1975	1,918	2.7	9,000	2.3
1976	2,038	6.3	9,500	5.6
1977	2,124	4.2	9,800	3.1
1978	2,207	3.9	10,100	3.1
1979	2,247	1.8	10,200	1.0
1980	2,286	1.7	10,300	1.5
1981	2,295	0	9,990	−3.0
1982	2,241	−2.4	9,960	−3.3
1983	2,310	3.1	9,860	2.1

Sources:
Monthly Energy Review, DOE/EIA, March 1984
Statistical Abstract of the U.S., 1982-83

All three levels have different economic characteristics. Most notable are the high fuel costs associated with peak-load production.

Studies both by the utility industry and the government indicate that the overall economic effects of distributed photovoltaic systems on utilities

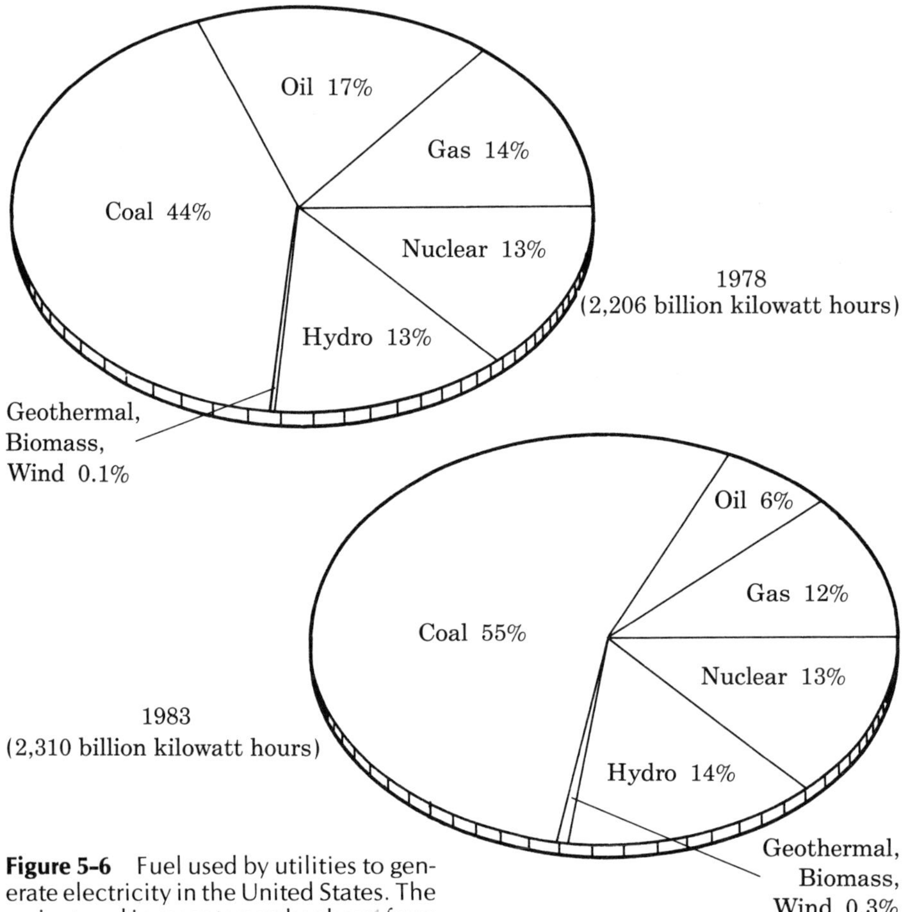

Figure 5-6 Fuel used by utilities to generate electricity in the United States. The major trend in recent years has been from oil to coal. (Source: DOE/EIA *Monthly Energy Review*)

will be very different at low (less than 10–20 percent) and high market penetrations.[13]

Low penetration by photovoltaic systems will have little effect on utility production costs. As more of these early photovoltaic systems come on line, they will displace some of the generating capacity needed during peak, intermediate, and base-load periods, allowing utilities to defer the construction of new generators. Photovoltaic use decreases coal, oil, gas, and nuclear fuel consumption (see Fig. 5-6) during all three load levels, replacing conventional generating capacity by somewhere between 0 and 50 percent of the rated capacity of the photovoltaic system. Photovoltaic systems have their highest output during midday, about the time when many Sunbelt utilities experience peak loads. To meet this peak, utilities presently must bring on line their older and usually smaller generators, which are less efficient and more expensive to operate. Peak loads are therefore characterized by higher fuel costs but lower capital costs because the older equipment has been paid for.

Only at higher penetration levels (that is, greater than about 20 percent of a utility's total output) would photovoltaics begin to replace base-load capacity; at such levels there would be a trade-off between saved fuel costs and the cost of storage systems. When the issue of storage comes into play, a new set of economic factors must be considered.

If storage systems are distributed, the residential owner may use any excess photovoltaic-derived power either to recharge the home storage system or to sell back to the utility company. The owner can be expected to recharge his storage system first and sell back later. If the utility buys back power during peak periods, the associated higher peak-load operating costs are offset.

If the utility itself maintains a storage system and draws upon it to cover peak loads, then distributed photovoltaic storage and central storage compete with each other.

At high-penetration levels, photovoltaics alone could supply the excess electricity needed during peaks and proceed further to recharge the utility's storage system. The utility's peak load will then occur at some other time of day, likely at the slightly lower levels around the current daytime peak, and could be served by system storage. Photovoltaic generation and system storage, then, become economically complementary. As can be seen in Figure 5-7, even limited storage capability can be important.

Until recently, it was thought that not before photovoltaic system prices fell to \$1.10-\$1.30 per peak Watt would utilities find it economic to replace conventional sources with their own solar arrays. This picture is now changing. Problems with finding sites for new conventional facilities, as well as climbing costs, are beginning to motivate some utilities to contemplate procuring photovoltaic units as an optional power source. When a utility needs a new generating plant of any sort, it is concerned with the cost of power from the new facility, not the cost from its combined existing generating capacity, most of which is bought and paid for. Especially in periods of high interest rates, utilities are concerned with the long lead time needed to construct conventional power plants and bring them on line—up to 12 years, in some cases. By contrast, photovoltaic systems can be placed in operation a

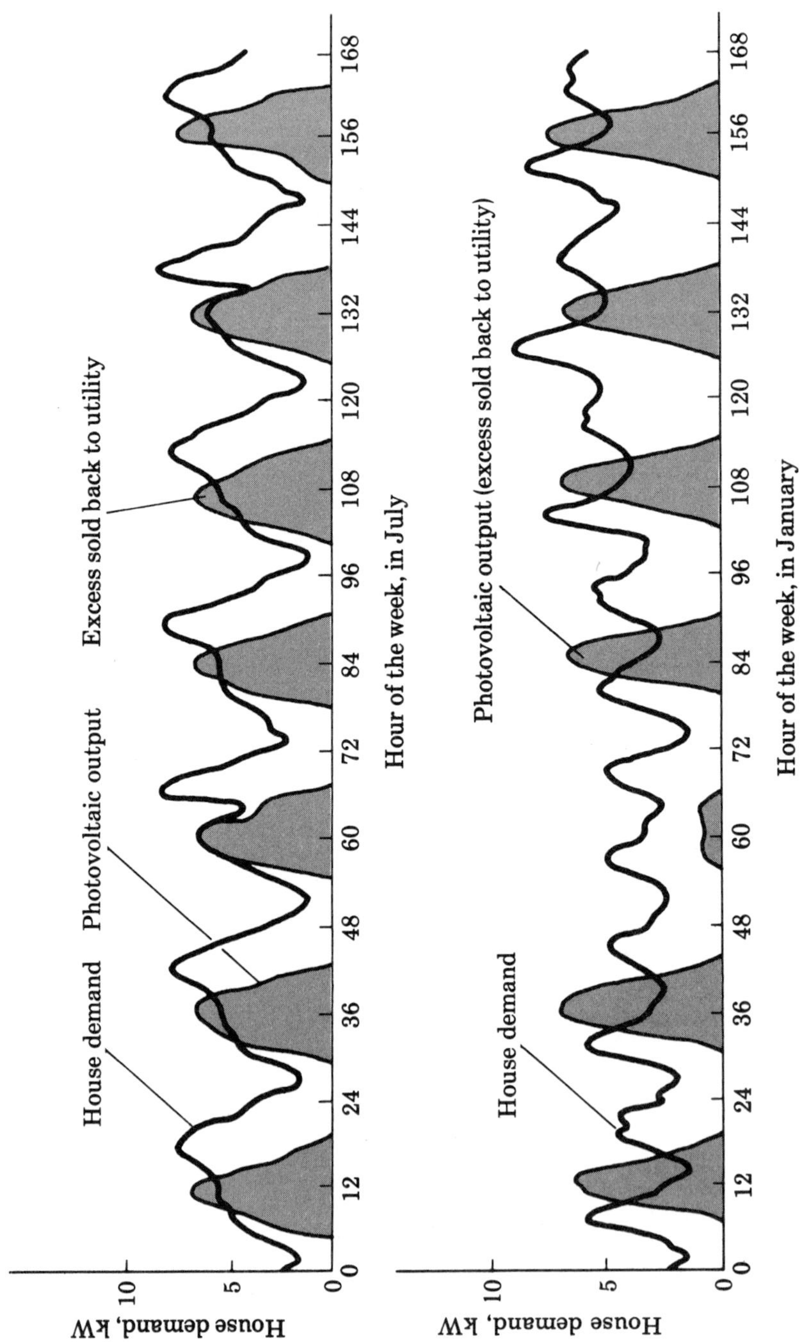
House demand
Photovoltaic output
Excess sold back to utility
House demand, kW
10
5
0
0
12
24
36
48
60
72
84
96
108
120
132
144
156
168
Hour of the week, in July
House demand
Photovoltaic output (excess sold back to utility)
House demand, kW
Hour of the week, in January

megawatt at a time, as they are completed. Thus, photovoltaics offers utilities an attractive new option in generating capacity.

The two most critical figures for a utility are the rate it is permitted to charge for the power it sells and the rate of return on investment it is permitted to earn.

Congress has already mandated, in the Public Utility Regulatory Policies Act (PURPA) of 1978 (Public Law 95-617), that utilities' rate structures may not discriminate against small producers of electricity, nor may these rates force other customers, in effect, to subsidize the same small producer. The Act requires a utility to buy power from distributed photovoltaic sources at a fair price, defined as the "avoided costs" to the utility—that is, it must pay an amount equal to what it has saved by not having to produce that electricity. Reasonable buyback rates, approaching 50 percent of the price of purchased electricity, will provide a strong boost for distributed photovoltaic systems. These rates will be established by law or set by public utility commissions, in accordance with the PURPA mandate. The California State Energy Commission now requires utilities to pay 100 percent of the utilities' own electric rates when they buy back excess power from customers who have solar installations.

Since buyback rates are based on the avoided costs to the utility, it is important that the utility show what generators it does *not* have to use for each hour it is buying power from homeowners. A study by the Argonne National Laboratory, done for the Office of Advanced Conservation Technology, U.S. Department of Energy, purports to show that residential photovoltaic systems would replace more base-load than peak-load power, and in some situations would even increase peak-load generation.[14] The issue is primarily one of accounting, which affects rate structures—a matter of future concern to both utilities and their photovoltaics customers. But it is still a peripheral issue. The true worth of photovoltaics should be measured in terms of broader economic questions than the cost of power foregone—questions such as the need for renewable energy sources as an alternative to rising fossil fuel prices, to the economic havoc that can result from overdependence on foreign fuel sources, and to the value of cleaning up the air we breathe.

Figure 5-7 Simulated residential electric power load and solar photovoltaics output (8-peak-kilowatt system), Phoenix, Arizona. Peak midday power production from a solar photovoltaic unit does not usually coincide with household demand. A few hours' storage capacity greatly increases the efficiency of the solar installation. Storage, though not essential, should be viewed as an integral part of the problem. (Courtesy: U.S. Dept. of Energy, with Aerospace Corp.)

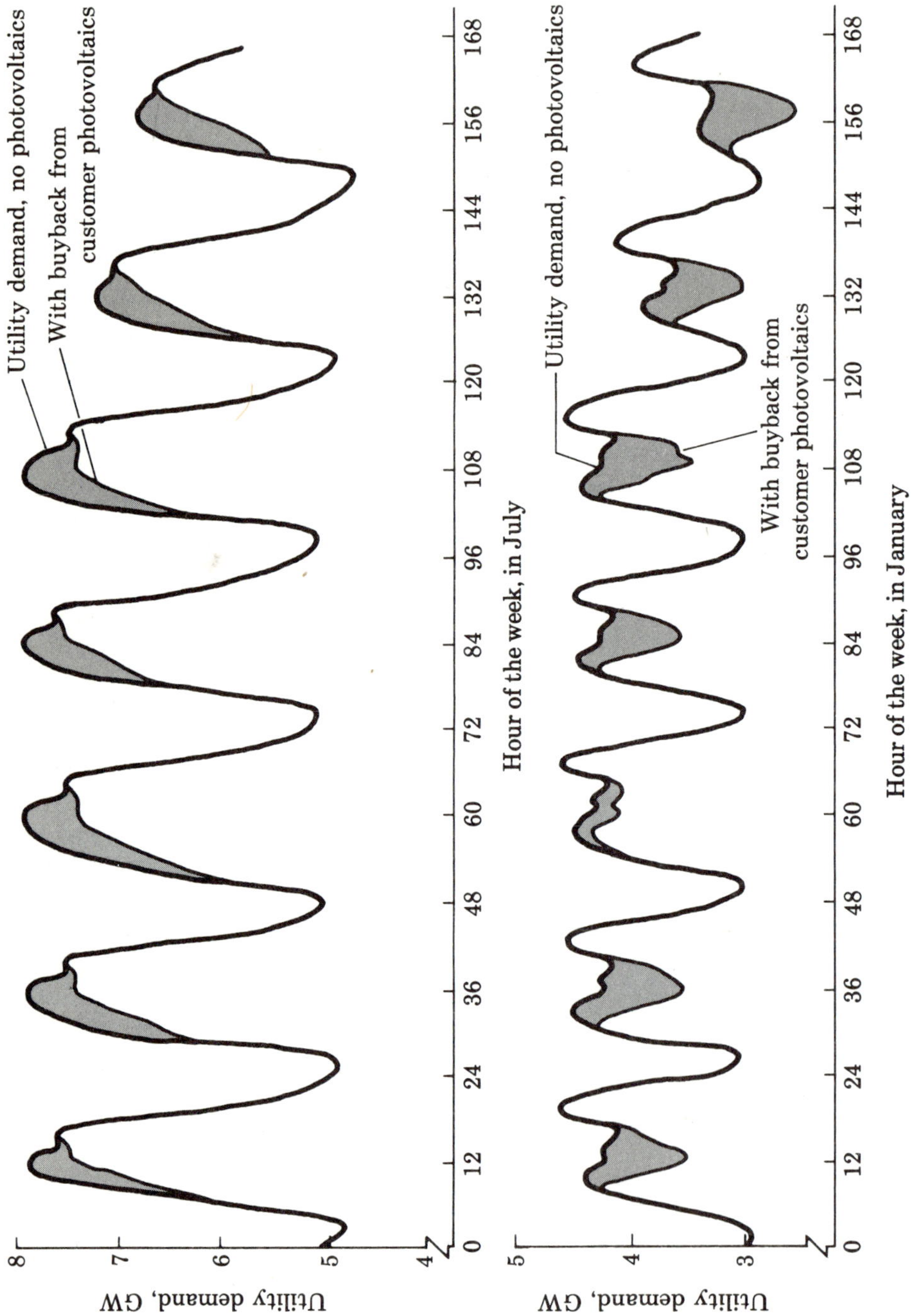
Utility demand, no photovoltaics
With buyback from customer photovoltaics
Utility demand, GW
8
7
6
5
4
0
12
24
36
48
60
72
84
96
108
120
132
144
156
168
Hour of the week, in July
Utility demand, no photovoltaics
With buyback from customer photovoltaics
Utility demand, GW
5
4
3
Hour of the week, in January

Thus far, the utilities' response to customer-owned photovoltaic systems has varied from full cooperation to disdain. The utility industry is not monolithic, and differing situations create differing responses. Some utilities have trouble getting adequate fuel supplies; others do not. Some have trouble with pollution regulations or finding appropriate sites for expansion; others do not. Some are good investments for prospective stockholders; others have chronic trouble raising funds.

A utility's decision to resist or embrace distributed photovoltaics depends on these factors and more. Some utilities, foreseeing a drop in absolute profits, may oppose the competition offered by these new generating sources that do not fall under their direct physical and economic control. Others, seeing the opportunity that distributed sources offer, may find much that is favorable.

Procurement of photovoltaic systems for central power-station use began in 1982. A total of 12 megawatts of generating capacity was funded by "third-party financing," an arrangement by which a utility offers to purchase electricity from a photovoltaic system owned by a group of investors. A typical time period is ten years, and the price to be paid for the power is the "avoided cost" to the utility of not having to produce power by other means. The investors, by procuring a solar photovoltaic system, can take advantage of the 10 percent investment tax credit, a 15 percent solar tax credit and accelerated (five-year) depreciation on the photovoltaic unit.

Under such an arrangement, a 1-megawatt unit was installed in 1982 near Hesperia, California. ARCO Solar supplied the equipment, and the power is sold to Southern California Edison. ARCO Solar also installed 6 megawatts at San Luis Obispo, California, in 1983. Similarly, another group of third-party investors procured 5 megawatts of capacity from United Energy Systems for a plant located at Borrego Springs, California. While it is difficult to predict how much photovoltaic capacity may be established in this way, the tax credits make systems that can be installed for $10 per peak watt or less economically attractive on a wide scale.

MARKET GROWTH AND GOVERNMENT POLICY

Government attitude is crucial because the future growth of photovoltaics depends as much on perceptions of this technology as on market events themselves.

Figure 5-8 Utility load profile with and without customer-owned photovoltaics for typical summer-peaking utility, Phoenix, Arizona. From the utility's standpoint, buying power from customers who own photovoltaic systems can alleviate peak-load requirements. A few hours' of storage capacity would further ease peak loading. (Courtesy: U.S. Dept. of Energy, with Aerospace Corp.)

A glance back over recent history is appropriate here. In the early 1970s, the National Science Foundation (NSF) began to support the photovoltaic effort with an eye to future possible uses on the earth. Its budget was then about $1-$2 million per year. When the Energy Research and Development Administration (ERDA) was established in 1974, most solar energy programs, including the NSF photovoltaic team, were transferred to the new agency. As the energy crisis sharpened, more attention was given each year to photovoltaics.

The DOE research, development, and demonstration budget in photovoltaics rose to $151 million in 1980. All along, the government's intention clearly was to help the technology to grow, primarily by assisting research and development, looking to the day when private industry would find this an economically successful way to produce electricity. Likewise the government intended to step out as soon as private industry assumes the initiative for development, production, and commercialization.

Under the pressure of OPEC oil shortages and consequent price rises, the government was compelled to do more than support research and development. Beginning in 1977, the federal government purchased solar cells in the marketplace in four successive "block buys" (large purchases) to boost private sector production and to lower unit costs, at the same time providing equipment for test applications and demonstrations. This program was quite successful, with module prices dropping from $40 per Watt of generating capacity to about $10 per Watt within three years—a striking example of how quickly the cost of supplying photovoltaic devices can come down with innovation and volume production.

Table 5-4 U.S. Department of Energy Purchases of Photovoltaic Modules in Relation to Total World Production, 1974-1983 (kilowatts)

	1974	*'75*	*'76*	*'77*	*'78*	*'79*	*'80*	*'81*	*'82*	*'83*
Private Sales	100	200	404	660	1,100	1,800	3,000	4,700	9,000	21,000
DOE Purchases	0	0	46	90	200	500	1,000	500	190	700
Total	100	200	450	750	1,300	2,300	4,000	5,200	9,190	21,700
DOE % of Total	0	0	10	12	15	22	25	10	2	3

As indicated in Table 5-4, government purchases of photovoltaic modules reached a high of 25 percent in 1980. The purchases were then phased out because industry was making quality products based on government-purchase specifications.

The federal government does have ample mechanisms to influence the course of a technology or an industry; these include direct grants, low-interest loans with or without government guarantees, direct procurements for government use, and various forms of subsidies to purchasers. In early 1980, photovoltaic systems were added to the classes of solar equipment for which a tax credit was allowed. The allowance was also raised to 40 percent of the first $10,000 for purchase or installation. Absolute amounts of further federal support for photovoltaics will depend on the perceived benefits to society. The mechanisms employed by the federal government will be based on many factors: the equities involved, ease of administration, possible time lags involved, and public perceptions of the acceptability and usefulness of such activities.

Congress has been a strong advocate of purchases under the Federal Photovoltaic Utilization Program (FPUP), which encouraged various agencies to buy and use photovoltaic systems, with funding provided through DOE. Nearly 4,000 separate projects were planned and funded.[15] Collectively, they generate about 750 kilowatts at peak power. Most are for small applications such as water pumping, communications, lighthouses, navigational aids, and power for residences for Park Rangers, the military, etc.

The record makes clear the value of government support of photovoltaics market development to the present time. It is arguable what form of support or inducement would be most effective and most equitable in the future: government purchases, as in FPUP; subsidies to users, as in the tax credit approach; or subsidies to producers, as through low-cost loans for equipment and fast tax write-offs, etc. In any event, government support of research and development is expected to continue at significant levels.

In addition to the FPUP program, a form of open-ended government purchase of research and development was initiated (Program Research and Development Announcement, PRDA). DOE has paid for competitive designs for a number of system evaluations and experiments and then contracted further with the winners to construct the systems. Seven PRDA projects are now in operation.

DOE has issued a detailed analysis of how much federal money might be required to meet the Congressional goals spelled out in the Photovoltaic Research, Development, and Demonstration Act of 1978. These goals include doubling the nation's photovoltaic generating capacity every year for the years 1978–88, and providing a total installed photovoltaic capacity of 4 million kilowatts (4 gigawatts) by 1988. In addition, systems costs are projected to fall to $1 per Watt by 1988, with the nation's private sector purchasing 90 percent of output. The Act authorizes, subject to appropriations on an annual basis, $1.5 billion to accomplish these ends. DOE has examined a number of options and concluded that within this $1.5 billion range, considerable progress could be made toward the goals of the Act. However, for assurance that the 1988 goals will be met, a budget of $3-$10 billion may be required. (By way of comparison, more than $100 billion has been spent to support development of commercial nuclear reactors.) Much of the additional support may come from private sector investments, and there are encouraging signs that this will happen. In 1984, we identified $82 million of private sector photovoltaic research and development in addition to the DOE budget. For this to continue, the government must be steadfast in its backing of the program.

Most of this money would go to building up production capacity by supporting early market demand. It is clear that the role of government in the initial phases of market expansion of photovoltaics is crucial.

INTERNATIONAL MARKETS

Government policy clearly is also crucial in shaping the international market. The United States currently has a large number of multilateral and bilateral agreements concerning solar energy. Foreign sales are projected to account for half of intermediate-term sales. DOE's International Photovoltaic Program Plan indicated that three major foreign intermediate-term markets are low-lift and medium-lift water pumping applications and electric power systems for remote villages. The total size estimates for these markets range from 16 to 315 peak megawatts per year. Foreign markets have a large potential for photovoltaic sales in the intermediate-term photovoltaic system price range of $3-$10 per peak Watt.[16] Another DOE report forecasting foreign market demands for photovoltaic systems arrived at similar conclusions indicating that foreign sales of communications and cathodic protection systems have the potential to grow more rapidly than U.S. sales in these markets.[17] Many multilateral and bilateral agreements address research, development, and demonstrations of photovoltaic technology. The majority mentioning photovoltaics call for exchange of research and development information. For example, the U.S.-Israel agreement includes a research and development project on luminescent solar concentrators. The U.S.-U.S.S.R. agreement includes an exchange of technical information on gallium arsenide and multiple-junction silicon cells. Such cooperative efforts may hasten maturity of photovoltaic technology.

Two important agreements include photovoltaic demonstrations. The largest project is the Saudi Solar Village Project, a 500-peak-kilowatt village power system serving two villages near Riyadh, the capital of Saudi Arabia.

The project cost approximately $17 million; equipment was procured from American firms.

In another agreement, a 20-peak-kilowatt experimental photovoltaic power system constructed beside a solar thermal experiment is part of a U.S.-Italy bilateral agreement. The U.S.-Italy agreement will also field a 5-kilowatt photovoltaic village system modeled after the Schuchuli Indian Village in Arizona.

It has been thought that the developing countries would be the major market for photovoltaics. Electricity for water pumping is a prime need, and photovoltaics is well suited to this since most of the countries lie in the Sunbelt and have ample insolation. Much of the land is semi-arid, but usually there is ample water within perhaps 50 meters (150 feet) of the surface.

In addition, there are an estimated 2 million villages in these countries with populations of 100–1,000 persons each. The people need a minimum amount of electricity for basic needs such as cooking, lighting, communications, crop grinding, and if possible for sewing and refrigeration.

Foreign sales are expected to grow steadily as the price of photovoltaics comes down. Even so, the United States market is also expected to expand rapidly in the 1980s, and the domestic market may quickly outdistance the foreign market in the intermediate term, 1985–90.

Principal purchasers in the foreign market are likely to be governments and international developers and the financing agencies that work with them.

HOW DO THINGS STAND NOW?

In the years 1973–80, the federal budget for photovoltaics research, development, and demonstration rose steadily, reaching $151 million in fiscal year 1980. The net effect was that, during the seven years following the Arab oil embargo of 1973 and the Cherry Hill conference,[18] a series of enlightened administrations and congresses provided a level of support of photovoltaics that, while not extravagant, was at least healthy; the effects were dramatic. In the years 1980–84, the federal support for commercialization was curtailed and the federal role shifted to supporting only high-risk research and development that promised long-term payoff.

The private sector is now allocating large amounts of risk capital—$100 million in 1983—to photovoltaic development. New plants with two-shift capacity, totaling 10–25 megawatts per year, are being built. (Annual sales in 1983 were only about 9 megawatts—one-fourth of the production capacity now under construction.) The industry appears to be planning an expansive and expensive growth to the billion-dollar-per-year level in the near future.

Since photovoltaics first began to attract interest because of the nation's space program, costs per peak Watt have already been reduced by two orders of magnitude, a factor of 100, and are now within a factor of 4 of being generally competitive with utility power sources in most parts of the United States.

It is too early to evaluate fully the effects of the sharp reduction in government support for photovoltaics under the Reagan administration, but the earlier momentum is clearly threatened. The single most important influence on the infant photovoltaics industry between today and the "pot-of-gold" market when installed systems will be universally available for $3 per peak Watt or less continues to be sustained, steady market growth. This can best be assured by continuing solar tax credits. Otherwise, manufacturers will not have enough confidence in the future of the market to build larger, more efficient plants and to invest in improved processes. Beyond that, growth will come from public realization that, far from being magical or exotic, photovoltaics is a mature and practical technology of today, which offers even more attractive benefits for tomorrow. Ultimately, public awareness of its benefits will provide the driving political and economic force needed to bring photovoltaics into widespread use.

NOTES

1. U.S. Department of Energy, Assistant Secretary for Energy Technology, Division of Solar Technology, *National Photovoltaics Program: Multi-Year Program Plan* (draft) (Washington, D.C.: U.S. Department of Energy, 1980).
2. The federal program is directed by the Photovoltaics Energy Technology Division of the U.S. Department of Energy. At field level, the research aspects are managed by the Solar Energy Research Institute (SERI), Golden, Colorado, which has an extremely well-equipped photovoltaics laboratory. SERI also coordinates contracted research at many universities and research centers.

   The Jet Propulsion Laboratory (JPL), Pasadena, California, provides field management for development, test, and demonstration activities, with principal support from the Sandia National Laboratories, Albuquerque, New Mexico. JPL oversees a number of contracts with various industrial firms.

   In each of these organizations, excellent technical management teams have been assembled to participate in planning and guiding their respective sub-programs.
3. Private communication, Royal Dutch Shell.
4. Louis Rosenblum, William J. Bifano, William A. Poley, and Larry R. Scudder, *Photovoltaic Village Power Application: Assessment of the Near-Term Market,* prepared for U.S. Department of Energy, Division of Solar Energy. DOE/NASA/1022-78/25, NASA TM-73893 (Cleveland: U.S. National Aeronautics and Space Administration Lewis Research Center, January 1978).
5. Private communication, Royal Dutch Shell. Proprietary study.

6. Louis Rosenblum et al., *Photovoltaic Village Power Application: Assessment of the Near-Term Market,* p. 21.
7. See Martin Wolf, "Historic Development of Photovoltaic Power Generation," photocopy (author affiliated with the University of Pennsylvania, Philadelphia, Pa. [ca. 1975]), Fig. 11; and U.S. Department of Energy, Assistant Secretary for Conservation and Solar Energy, *Federal Policies to Promote the Widespread Utilization of Photovoltaic Systems,* DOE/CS-0114/2 (Washington, D.C.: U.S. Department of Energy, 1980).
8. Press release, Solarex Corporation, 23 June 1980, Rockville, Md.
9. *Federal Policies to Promote the Widespread Utilization,* Vol. 2 (preliminary), pp. 6–8.
10. *Federal Policies to Promote Widespread Utilization,* Vol. 2, pp. 12–48.
11. Letter to the editor, *New York Times,* 16 May 1984.
12. Michael Totten, Director of Public Citizens Critical Mass Energy Project, letter to the editor, *New York Times,* 29 May 1984.
13. *Federal Policies to Promote Widespread Utilization,* Vol. 2, pp. 2-10–2-11.
14. Mueller, R. O., B. K. Cha, R. F. Giese, "Solar Photovoltaic Power Systems: Will They Reduce Utility Peaking Requirements?" *Science* 213 (10 July 1981), pp. 211–213.
15. *FPUP Activities,* Vol. 4 (Pasadena, Calif.: Jet Propulsion Laboratory, April 1982). *Federal Photovoltaic Utilization Program Annual Report for Fiscal Year 1981* (Washington, D.C.: U.S. Department of Energy, December 1981).
16. U.S. Department of Energy, *International Photovoltaic Program Plan: Report to the Congress in Accordance with the Solar Photovoltaic Energy Research, Development and Demonstration Act of 1978,* Vol. 1 (Washington, D.C.: U.S. Department of Energy, 1979), p. 18.
17. U.S. Department of Energy, *Export Potential for Photovoltaic Systems: Preliminary Report,* prepared by Battelle Pacific Northwest Laboratories, DOE/CS-0078 (Washington, D.C.: U.S. Department of Energy, 1979).
18. A meeting of scientists, engineers, and others active in the development of photovoltaics from industry, government, and academia, held at Cherry Hill, New Jersey, in 1973 under sponsorship of the National Science Foundation. The meeting was a landmark in the movement to make photovoltaics practical for use on the earth.

FURTHER READING

Application Analysis and Photovoltaic System Conceptual Designs for Service/Commercial/Institutional/and Industrial Sectors. Final report. SAND-79-7020. Durham, N.C.: Research Triangle Institute, 1978.

Conceptual Design and System Analysis of Photovoltaic Systems. ALO/3686. Philadelphia, Pa.: General Electric Co., March 1977.

Federal Policies to Promote the Widespread Utilization of Photovoltaic Systems. DOE/CS-0114. Washington, D.C.: U.S. Department of Energy, February 1980.

Kapany, N. S., ed. *Finance and Utilization of Solar Energy.* Santa Cruz, Calif.: Center For Innovation and Entrepreneurial Development, University of California, 1983.

Photovoltaic Power Systems Market Identification and Analysis. Volume I and II. HCP/M2533-01. Albuquerque, N.M.: BDM Corporation, 1978.

Pittman, P. F. *Conceptual Design and Systems Analysis of Photovoltaic Power Systems.* Final report. ALO/2744-13. Pittsburgh, Pa.: Westinghouse Electric Corp., Research and Development Center, April 1977.

Rattin, E. J. *Mission Analysis of Photovoltaic Solar Energy Conversion. Survey of Near-term (1976–1985). Civilian Applications in the United States.* SAN-1101-PA8-1/2. El Segundo, Calif.: Aerospace Corp., 1977.

Walton, A. L., and Warren, E. H. Jr. *The Solar Alternative: An Economic Perspective.* Englewood Cliffs, N.J.: Prentice-Hall, 1982.

CHAPTER 6
THE POTENTIAL IMPACT OF PHOTOVOLTAIC TECHNOLOGY

There is every sign that our society will increasingly choose electricity as the preferred form of energy. Electricity is clean. It does not have to be hauled about. It is extraordinarily versatile as it can be converted easily into light, heat, or motion. It can be used with great precision—in a computer, a television set, or a surgical device. Moreover, the distribution system, representing an investment of $1 trillion, is already in place. In every sense, electrical energy is high-quality energy.

In the United States, we now produce and use about 260 gigawatt-years of electricity each year. (A gigawatt-year is the amount provided by a 1,000-megawatt plant operating without interruption for one year). Despite more efficient electrical equipment and other conservation measures of the last few years, the use of electricity continues to increase at about 3 percent a year. At this rate, some 400 gigawatt-years would be consumed in the year 2000. Even stringent conservation is unlikely to reduce this figure significantly. About 200 new generating plants of 1,000 megawatts each would have to be built before then to accommodate the increase.[1] Where the additional energy will come from is an open question, but it is conceivable that photovoltaics could provide most of it.

It is important to recognize some singular aspects of massive photovoltaic deployment. For instance, the cost of photovoltaics is almost entirely attributable to the purchase and installation of the system; it is a one-time expense. There are no fuel costs, and maintenance costs are minimal. Most of the expense of the changeover to this new energy source would be borne by the private sector as a normal part of new construction costs or in connection with the maintenance of existing structures. Such costs would in essence be attributable to higher quality of living. Consider: only a few decades ago, most new housing did not have central heating. Now central heating is taken for granted, and the same goes for electricity. Our economic system is able to provide these amenities and many, many more attributes of modern living. Some people earn their living providing them, and all of us enjoy the results. So it will be with photovoltaic solar energy.

The primary issues are whether technology can provide the capability at reasonable expense in terms of the human and material resources needed and what the social and economic consequences of widespread use of photovoltaics may be.

MATERIAL AND RESOURCE CONSTRAINTS IN THE DEVELOPMENT OF PHOTOVOLTAICS

For any technology to expand rapidly into the marketplace, there must be an adequate supply of materials at costs reasonable to the purchaser, or the technology must be amenable to changes that bypass critical shortages.

In only a few instances are special materials for specific cell types in critical supply.[2] Polycrystalline thin-film cells will require increased production of trichlorosilane. For amorphous silicon cells, indium is the critical material in present cell designs. Cadmium will be the pacing material for cadmium sulfide-copper sulfide cells, but proven world reserves of 700,000 metric tons indicate no shortage.

For gallium arsenide concentrator cells, availability of gallium, germanium, and arsenic may restrain large-scale production. But concentrators require small cell area, and such systems are useful only in regions of high direct insolation.

No material shortages are anticipated in any of the silicon designs because there is enough lead time to provide needed production facilities. Purified silicon will be required in greatly increased quantities. Several production processes are available, and two companies, Union Carbide and Solarex, are in early production. Both claim their new plants will be able to reduce the price of silicon.

The diversity of cell designs and concepts assures a wide choice of materials. Material availability and cost will be only two factors among many that determine which concepts find their way into the marketplace. The advent of thin-film cells (less than 100 micrometers thick) and automated production processes will tend to make supplies ever less critical. Further, no single design is likely to corner the market, and a mixture of cell types will further offset any shortage.

Widespread use of photovoltaics will have its greatest impact on ordinary structural material: iron, steel, aluminum, glass, and copper. Until advanced batteries come on the market, lead may be in increased demand. Indeed, of all the materials involved, it appears that glass will be the pacing item, since it will be the material most used as the weather surface for photovoltaic modules. A few years' lead time for expanding production will prevent a shortage of glass from becoming a problem.

PHOTOVOLTAICS AND THE ENVIRONMENT

Photovoltaic cells are sealed units and no by-products are created in normal operation. However, it is possible for a cell to fail in a way that produces a local hot spot. Under extraordinary conditions, short circuits could occur and cause sparks, with consequent risk of fire. Electrical protective devices are available and would be incorporated in systems to sense such abnormalities and shunt the current past the troubled part. In general, electrical safety can be assured through normal code practices regarding insulation, lightning protection, safety fusing, and related matters. With reasonable attention to such aspects of system design and installation, photovoltaic systems should prove fully as acceptable from an operational standpoint as, for example, home television sets.

In the longer future, the partial shading of significant areas of land by massive photovoltaic arrays merits consideration. Shading can be greatly alleviated by the use of roof areas and south-facing walls for arrays. Increased cell efficiency will give most homeowners adequate array space on their dwellings. Moreover, land that is subject to partial shading can be used for many purposes. Horticulturists use this principle. In many areas of the world, additional shade would be a distinct benefit. The matter is not considered serious because the total land area needed in most countries will be small compared to what is available.

From the standpoint of material hazards, there are three broad areas of concern: the release of toxic gases due to system malfunction; exposure of industry workers to toxic gases and dusts; and solid waste disposal, or release of gases, during photovoltaic material mining, manufacture, or disposal. All of the materials involved are well known to industry, and normal Occupation Safety and Health Administration (OSHA) practice should provide adequate safety. Most of the potential hazards can be mitigated with some research, normal OSHA practice, and genuine concern for potential problems.

Release of toxic gases during system operation and malfunction, for instance, could lead to the possibility of asphyxiation from the toxic products of material decomposition. Materials such as electrical insulation plastics have caused death in fires involving electrical equipment other than photovoltaic systems. Some polymer materials used in concentrating optics (e.g., methyl methacrylate and vinyl chloride) release noxious fumes in a fire. Arsenic in gallium arsenide cells could vaporize in a fire and oxidize to a highly toxic arsenic trioxide. The decomposition and vaporization of cadmium sulfide at very high temperatures could lead to toxic fumes.

Exposure of workers in the photovoltaic industry to toxic gases and dusts will require monitoring and control as for any industry. The materials being considered for photovoltaic energy systems—silicon, gallium arsenide, copper sulfide, copper oxide, tellurium compounds, etc.—are already produced in major quantities by the $10 billion semiconductor industry, which has a good record in meeting worker-safety standards.

Approved health and safety codes from such industries as plastics, glass, and chemicals should be directly transferable to photovoltaic manufacturing operations. Photovoltaic cell manufacture presents no unique hazards and should be much less hazardous than coal mining.

Other matters that will require attention include exposure to silicon dust in cell preparation, phosphine and boron trichloride in cell doping, and cadmium compounds during material production; inhalation of arsenic compounds during cell fabrication; high-pressure release of arsenic trioxide in preparing gallium arsenide; and the handling of similar toxic agents. It will be important that toxicological research be expanded if new species and levels of toxic materials are generated. The industry, government agencies, and environmental researchers are beginning to perform these studies.

The disposal of wastes derived from photovoltaic systems production will require normal monitoring and control as production is expanded. For example, gallium extraction yields mercury and large amounts of alumina sludge.

Widespread application of any new energy technology can expose a large number of users and production workers to health hazards and create waste problems. However, photovoltaic systems are expected to enjoy extensive deployment without incurring exceptional environmental problems,[3] and the use of photovoltaics in lieu of expanded fossil and nuclear facilities will clearly be a net gain from the standpoint of protecting the environment. Dr. Louis G. Stang of the medical department at the Brookhaven National Laboratory, an authority on the environmental health aspects of photovoltaics, says: "The conclusion that can be drawn thus far is that photovoltaic technology is one of the cleanest of all energy options and certainly deserving of vigorous development."

EFFECT ON EMPLOYMENT

The employment potential of photovoltaics can be estimated rather simply. It takes 50 gigawatts (50 billion Watts) of photovoltaic generating capacity operating for one year to produce one quad of energy in a typical

U.S. location of medium insolation. At a system cost of $2 per Watt, which will be reached in the mid-1990s, 50 gigawatts would have a price tag of $100 billion.

The labor cost of manufactured products like this is about one-fourth the sales price of the product, in this case $25 billion. Let's assume that the average wage for a worker in the industry is $20,000 per year. Then $25 billion divided by $20,000 per person-year gives 1,250,000 worker-years required to fabricate 50 gigawatts of photovoltaic systems. Installation and service could add perhaps 20 percent, or another 250,000, making a total of 1,500,000 worker-years. This is the number of jobs required to produce the systems and install them and service them the first year. (The maintenance function would extend over the life of the installation, which could be 20 years.)

Widespread adoption of the various solar technologies would create an enormous number of jobs, according to the Council on Environmental Quality.[4] A general indication of the distribution of skills likely to be needed is given in Table 6-1.[5]

Table 6-1 Jobs Created through Solar Technology

Type	*Percent*
Scientists, engineers, architects, and managers (about 2,000 professionals presently are working on photovoltaics)	7
Foremen, technicians, and semi-skilled production operators	27
Manufacture of piece parts	8
Sales and merchandising	5
Manufacture of power conditioning electrical products	8
Installers: electrical, roofing, and plumbing trades people	18
Maintenance and service people	27
Total	100

Before the end of the century, a labor force of several million people will be in new jobs, not jobs taken away from present energy industries. Coal miners and others working today will not lose their jobs to the photovoltaics industry. The *increase* in energy will be provided by people working in photovoltaic production facilities and on installations.

IMPACT ON MUNICIPALITIES

The adoption of photovoltaics will bring some special opportunities and also some new responsibilities for municipalities and other local jurisdictions. It will be most important to assure that installation, maintenance, and service organizations are qualified through training and are fair in their dealings with the public. As the technology comes into the marketplace, entrepreneurs will seek to establish themselves. Some will be better qualified than others, and local government and better business bureaus must prepare to protect the public.

Aesthetics may become a concern, and how local government will face this issue is an open question. It will be not so much a new issue as a new twist to an old one. Photovoltaic arrays will not be a problem on new construction because they will be roof or wall panels or possibly imitation shingles designed into the building. They will be removable so that they can be replaced in event of damage or when their conversion efficiency finally, after 20 years or more, declines to a level that replacement is economically advantageous.

Arrays added to existing structures may not be as visually harmonious, especially in the early years, but architects and installers will learn from experience how to protect basic building lines while obtaining adequate exposure to the sun. Arrays will soon appear on the market in the form of siding, resembling shutters, and in all types of roofing patterns. This variety will help make photovoltaic arrays adaptable to almost any kind of structure.

It is no simple task to predict how homeowners will feel about owning photovoltaic systems. Arrays will have to be mounted in full view on their houses and will need periodic hosing down and occasional repair. There will be associated control devices and perhaps a bank of storage batteries. Most of all, owners will have to pay for the systems in advance or through financing, trusting that the equipment will operate at least long enough to break even with the cost of power from a utility—a matter of several years.

Large arrays of solar cells may at first look out of place. A few years ago, a Washington official, recalling publicity accorded the proposed new radar-invisible bomber and the $20 billion which Congress voted for the synfuels effort, wryly remarked that the sure road to success for photovoltaics would be to call solar cells "syncells" and make them invisible.

But soon photovoltaic arrays will become status symbols, and people will not want them to be entirely inconspicuous. In the early days of television, the proud owner of a new TV set was happy to have an antenna anchored high on the chimney. Eventually, photovoltaics will assume the role of a necessity. Architects will be offered a new challenge in blending arrays into structures harmoniously.

In time, sun rights will become a concern, and local government will enact ordinances to protect the property owner from loss of direct sunshine on his or her array. The owner will be free to sell sun rights, that is, an easement. This aspect of law will gradually be thrashed out in the legislatures, municipalities, and courts.

Finally, municipalities, especially larger urban areas, can look forward with optimism to ample supplies of electric power from nonpolluting photovoltaic sources, enabling them to reduce air pollution especially when the electric auto appears on the scene.

PHOTOVOLTAICS AND NET ENERGY

One key question about the effectivenss of a new energy option is how long must a renewable energy generator make electricity before the energy produced equals the energy required to manufcture the generator? This is called energy payback time. There is much confusion as to precisely how to do net energy analysis and, therefore, how to determine the payback times for various options.[6,7] Most previous analyses indicated that photovoltaic technology had a payback time of four years. That is, it would take four years before a photovoltaic system generated enough electricity to equal the energy needed to manufacture it.

However, new processes for photovoltaic production now assure both lower costs and shorter energy payback times, typically less than one year. Some of the advanced films like amorphous silicon and cadmium telluride can achieve energy payback in a matter of weeks.

In order to determine payback, the energy content of all raw materials must be added to the energy used in manufacturing, shipping, and installation. With payback times of less than one year, rapid market penetration can occur without greatly affecting the total energy consumed. In a detailed analysis prepared by Solarex, the concept of the solar breeder was proposed. Simply stated, a photovoltaic array would power a plant that would produce photovoltaic arrays. The power capacity of the arrays produced would offset the power used by the plant. It is now clear that production of photovoltaic systems can be undertaken on a massive scale without significant net energy loss to society.[8]

It should also be noted that once a photovoltaic generator with the ability to save one barrel of oil is installed, then that barrel can be saved again and again without any investment other than maintenance for the life of the installation, possibly 20 years or longer. That is, the energy capacity of photovoltaic systems is equivalent to having a proven reserve 20 times larger than the energy rating. For example, if the nation had an installed photovoltaic capacity that could produce in one year an amount of electrical energy equivalent to that obtainable from 10 quads of oil, it would be like having

200 quads of proven oil reserves. In addition, the "reserves" would be there at a guaranteed fixed price. This is the great value of a renewable energy option like photovoltaics with its quick (less than one year) energy payback time. In an era when rampant price increases for oil and other energy sources may recur, it would be comforting to have an installed photovoltaic energy system that will produce electricity at a fixed price for the life of the system.

PHOTOVOLTAICS AND INFLATION

Because energy is a fundamental ingredient in everything our society makes or does, changes in the cost of energy have a major effect on the economy. The most dramatic recent example was the drastic surge of inflation that followed the 1973 oil embargo and subsequent escalation in the price of oil. Former Treasury Secretary G. William Miller, also former chairman of the Federal Reserve, said that about one-third of the inflation was attributable directly to oil price escalation, and the "indirect effects would be considerably more."[9] The surge in oil prices was quickly translated into higher prices for all forms of energy, regardless of kind or source.

Subsequently, we learned that the reverse could also happen. After several years of prodigious effort to reduce consumption of foreign oil, helped along by a severe recession, which was induced to curb the inflationary spiral, an oil glut ensued. In the early 1980s, stable oil prices made two years of rapid economic recovery possible, free from inflationary pressures.

Dependence on finite energy sources subjects us to an economic roller coaster with all the human frustration and social malady entailed. To the extent that our country continues to be dependent on energy sources of finite supply, and especially if they are under concentrated ownership, we are subject to further serious episodes of economic dislocation. As nations become ever more closely tied together, continued and increasing dependence on oil, coal, gas, and nuclear energy is a major destabilizing influence in the world. This state of affairs, so sharply inscribed in recent memory, can only become more intense until we succeed in making the transition to renewable energy sources.

The world cries out for stability while it works out other pressing social and international problems. New energy sources that are abundant and economically available are desperately needed. Due to its simplicity, its favorable and steadily improving economics, and its vast technical promise, photovoltaics is now clearly the leading candidate for an energy source to free us from future inflationary-deflationary economic spirals, which result from, or are exacerbated by, the manipulation of prices in conventional energy supplies.

A significant shortcoming of all present major energy sources (hydroelectric power is a relatively small source) is that they directly contribute to the expectation of further inflation in prices. Because of the circumstances surrounding their production and use, oil, coal, natural gas, and nuclear energy may even be regarded as fundamental contributors to inflation. Because the supplies are obviously limited and being steadily depleted, they are inflationary energy sources.

Photovoltaics can be applied in ways that will relieve U.S. dependence on foreign oil. This should be the first objective. Certainly photovoltaics will not cause any reduction in the amount of coal being mined now, not within the lifetime of anyone now living. Photovoltaics *will* make it unnecessary to triple coal production, which some would now have us believe is the only practical alternative left.[10] Similarly, the advent of photovoltaics will not cause nuclear reactors now in operation to be abandoned. But photovoltaics can make it unnecessary to build more nuclear reactors or to turn to the breeder reactor, which society is clearly loath to do. No matter how fast photovoltaics comes on, present energy supplies—other than foreign oil—and the jobs associated with them are not likely to be curtailed. It is the additional supplies of the future that now are at issue, and there photovoltaics will make a difference.

PHOTOVOLTAICS OVERSEAS

It would be derelict to forget the millions who live in poverty in other lands. It is to them that the oil price hikes have dealt the severest blow, as the Brandt Commission pointed out.[11] In urban areas of developing countries, electricity costs in 1977 were $0.50 per kilowatt hour. In rural areas, power from diesel generator sets was $1 per kilowatt hour or more and $12 per kilowatt hour from primary batteries. In the years since, matters have grown steadily worse. Photovoltaic technology can bring hope to many nations that can no longer afford to import oil for the 15 million diesel electric power generators now in use.[12] The spillover effect from the application of the technology in developing countries could have beneficial long-term political results potentially as important to the United States as solving the foreign oil problem today. Providing electrical power at the village level in the underdeveloped world can bring hope. It can also open the door to education, which is essential for beginning to deal with such seemingly intractable problems as excessive population and consumption growth rates.

But whether it should be or not, humanitarian concern is not the only motivation, perhaps not even the primary reason, for present interest in transferring this new technology into the developing world. From a pragmatic point of view, this technology needs to be applied in places where it now makes sense economically, in order to build up the production base and thereby bring prices down. If Village X is about to buy a diesel generator for the first time, or Village Y about to replace its worn-out generating equipment, and a photovoltaic system is competitive, it is in both their interest and ours that they consider photovoltaics. Already, photovoltaic systems are directly competitive with diesel-electric power in countries that must import oil, if the cost of oil over the life of the system is included. As the price of oil continues to climb and the price of photovoltaics continues to drop, the scales will steadily tip toward photovoltaics. In the near future, all diesel generator sets will be replaced with photovoltaics as they wear out (unless a country has it own indigenous oil supply). One of the world's largest manufacturers of diesel-electric generators has decided to add a line of photovoltaic equipment to its offerings.[13]

A number of federal agencies are involved in demonstrating photovoltaic systems in other nations and facilitating their widespread use. Activities have included both system experiments abroad and system field tests in the United States under conditions similar to those in developing nations. In addition, some resources already have been devoted to the development of balance-of-system components for use in international applications.

The Agency for International Development (AID) and DOE have been responsible for photovoltaic projects including the refrigeration of medical supplies, water pumping applications, and general village power systems. A village system for Tangaye, in Upper Volta, is working well, for example; the grain-grinding experiment has increased productivity so that the village is now a net seller of food. The use of photovoltaic electricity to pump fresh water, as well as for lighting, communication, and food preparation and refrigeration can be a major factor in the lives of several hundred million people.

A final word of caution is in order. With photovoltaic technology loosed on the world, our civilization will never be the same. History has shown that the ramifications of any new technology can be unexpected. But such speculation is hardly reason to hesitate. We have to move ahead one step at a time, choosing as best we can, or perish. By any measure, the present outlook for photovoltaics is positive.

OTHER ISSUES

On energy matters, U.S. society is segmented in terms of ownership, uses, employment, living habits, even differences in climate from one region to another. These things pit different parts of society against each other in endless struggle. But ideologically there are two camps. Those persons and organizations that are users of energy are spearheaded by populist leaders who are committed to such social goals as a better physical environment and more equitable access to and control over energy sources and energy-distribution enterprises. Confronting this group are the energy-producing and -supplying organizations, corporations, and their owners (and those who see matters as they do). The struggle between these two camps is nothing new. For years it has been punctuated by disputes and brawls until hardly an eyebrow is lifted over another contest about, say, public versus private ownership of an electrical utility, or one more court test between "dams" versus "parks" groups.

Nevertheless, the energy crisis has intensified the usual interplay between the two encampments as never before. The march of events focuses unresolved issues in a way that produces severe strains on our whole socioeconomic system. For example, Amory Lovins, leading proponent of appropriate technology, says of the utilities that they are "very powerful and unenlightened. Even if under public ownership, they won't become benign. In Britain, they became less responsive . . . messing around with organizational form doesn't help."[14] He advocates energy technologies that will put control of energy supplies directly in the hands of the people. Even in places such as Minnesota, where all electric utilities are publicly owned, the gap between the managers of central power stations at one end of the line and the users at the other sometimes seems unbridgeable. The advent of photovoltaics, which can place at least a portion of the generating equipment in the hands of the user, offers hope that a new relationship will develop between utilities and their customers.

We are entering a truly fascinating era in the history of electric power generation and distribution. Since 1882, when Edison ran wires down the street from the world's first commercial central station generator in New York City, the trend has been all one way—toward bigger generators and longer lines. Photovoltaics offers the first real opportunity to reverse that trend.

The utilities are now having serious problems keeping up with demand. The famous blackout of New York City on November 9, 1965, brought the matter to public attention in dramatic fashion. The energy crisis intesified the utilities' woes. Not only must they meet the growing load, but they must do it in the face of fuel scarcities. Worst of all, spiraling fuel costs compel utilities to request repeated rate hikes, which further infuriates the public.

The executive director of the National Electric Reliability Council has pointed out that the utilities, which collectively burn 1.7 million barrels of oil a day, have been urged to scale this down to 1.5 million barrels per day by 1989.[15] Even this small cutback, he says, will take a prodigious effort. "The financial condition of the electric power industry is the worst it has ever been," says Gordon Carey, vice chairman of Commonwealth Edison of Chicago; this makes it virtually impossible to enlarge capacity or switch energy sources.[16] The utilities, it is clear, are caught in a blind alley.

Under these circumstances, photovoltaics could be a boon to the utilities as well as to the public. The utilities are offered an alternative power source that can be located anywhere in the system because it is modular in nature, has near-zero pollution, and is likely to be popular with the public. Homeowners can, if they wish, provide their own electricity with a photovoltaic system and be freed from total dependence on the utility. Utilities are relieved from the steady pressure to provide new generating capacity, and they can maintain earnings by buying electricity from homeowners during daylight hours at reduced rates and selling it back at night at regular rates. Utilities can make money as distributors of power as well as by selling power they produce. From the public standpoint, the tremendous distribution network—paid for by the users—is an asset not to be scorned.

How the utilities and the public agencies that regulate them respond to photovoltaics will depend on their understanding of the technology and what it offers, and a willingness on all sides to try new approaches. The path will not be easy for the utilities. Photovoltaics involves a new technology with which they are not familiar; new forms of ownership, as discussed in the preceding chapter; and modifications of relationships with their customers. But to those companies able to exploit it effectively, the rewards could be appreciable. Fortunately, utilities in many parts of the country are becoming interested, and several are actively engaged in projects of their own. In fact, photovoltaics provides options that we have never had before—not since the dawn of commercial electric power. Lawmakers, regulators, utilities, and the consuming public will be challenged to break out of traditional thinking and hidebound customs and seek new approaches that take advantage of the opportunities photovoltaics now offers.

An issue of social equity arises in the early stages of commercialization. Initially, the most likely residential purchaser is in a high-income bracket. The 40 percent tax credit (to a maximum of $4,000) offers the wealthier members of society greater opportunity to procure photovoltaic systems. At first glance, this may appear grossly inequitable. But if someone must "break the ice," why not the more affluent? It is no different from any other new technology; the wealthy were the first to purchase home air conditioning, the private auto, electricity, and television. Someone must be first to buy photovoltaics so that the industry will grow and prices fall due to mass production and technological advances. Whoever is the first to buy photovoltaics is saving oil or other fuels, reducing air pollution, and above all bringing photovoltaics one day closer for the rest of us.

In something so fundamental as energy, political authority can merely promote an attitude of reasonableness among the contending parties to assure that society continues to function smoothly while the people sort out their priorities in the marketplace, political arena, and their private lives. It is the people who finally must decide what energy future they want and why. If adequately informed, Americans will reach the right decisions and in a reasonably harmonious and equitable way.

THE TOTAL MARKET: HOW LARGE AND HOW SOON?

Will society adopt this new technology and apply it rapidly, or will photovoltaics be allowed to languish, coming slowly into use in some far-off future? What will be the size and timing of the total market? These questions follow a discussion of societal issues because too often the future of new ventures has been projected on the basis of purely economic factors leading to erroneous conclusions. Nuclear power looked good, economically, until public fears over safety matters were reflected in escalating design and construction costs. Now the public is shying away from it on grounds both of poor economics and adverse safety factors. One wonders if coal may in time suffer a similar setback.

Concerns about market potential have spawned a series of studies on the future of photovoltaics. Unfortunately, each has focused on some special issue or aspect of the subject or was motivated by some unique concern. The issue of total market size and timing has never been properly addressed. An extensive study by BDM focused on present uses and near- and intermediate-term applications only, involving small units for use at isolated sites.[17] It did not consider major applications to save energy. The SERI Venture Analysis examined whether, and under what circumstances, a proposed $380 million, eight-year "market pull" initiative to promote production and reduce costs would be justified; a series of market scenarios assumed for purpose of analysis were of incidental importance in themselves and admittedly involved great uncertainities. The content and objectives of earlier studies are summarized in the SERI effort.[18]

The history of government forecasts of the commercial potential of photovoltaics is a classic example of the political process at work as federal authorities consider the possible widespread impacts of a pervasive new technology. In April 1978, the Council on Environmental Quality, a White House policy advisory group, officially a part of the Executive Office of the

President, produced its assessment on solar energy.[19] The Council projected that photovoltaics could provide 2–8 quads per year by the year 2000 and as much as 10–30 quads per year by 2020, as indicated in Table 6-2. (One quad corresponds to 50 gigawatts of photovoltaic capacity operating for one year.) The high mark given photovoltaics reflected the confidence in its technical growth potential as well as in its basic simplicity and reliability.

Table 6-2 Estimates by the President's Council on Environmental Quality for Maximum Renewable Energy Sources Contribution to U.S. Energy Supply under Conditions of Accelerated Development (April 1978). Units represent quads of displaced fossil fuel per year.

Technology	*1977*	*2000**	*2020**
Heating and cooling (active and passive)	Small	2–4	5–10
Thermal electric	None	0–2	5–10
Intermediate temperature systems	None	2–5	5–15
Photovoltaics	Small	2–8	10–30
Biomass	1.3	3–5	5–10
Wind	Small	4–8	8–12
Hydropower	3	4–6	4–6
Ocean thermal energy conversion	None	1–3	5–10

*The estimates in these columns are not strictly additive. The various solar electric technologies will be competing with one another, and their actual total contributions will be less than the sum of their individual contributions.

A different kind of study was the Domestic Policy Review (DPR) of Solar Energy conducted in 1978 at White House direction.[20] The DPR was a thinly veiled effort by the Carter administration to turn aside political pressure from a collection of solar enthusiasts. Various activist groups, which included serious students of the energy crisis as well as anti-nuclear, anti-big business, consumer protectionist, environmentalist, and similar elements, had rallied around the solar energy banner. Under the brilliant leadership of Denis Hayes, a highly publicized event called Sun Day was celebrated on May 3, 1978, to promote interest in solar energy. Speaking that day from the top of a mountain that was to become the home of the Solar Energy Research Institute, on the western outskirts of Denver, President Carter announced that the DPR would be held under the aegis of the Secretary of Energy and 17 other department and agency heads. Whenever so many government units are deliberately included in a study, it is likely to be a stalling operation rather than an agent for change.

The DPR addressed all the renewable energy options, not photovoltaics alone. The outcome, announced more than a year later by the President at the unveiling of a solar installation on the roof of the White House, called for 20 percent of the national energy consumption to come from solar (renewables) by the year 2000 (19 quads of a projected total energy demand of 95 quads).[21,22] The portion foreseen to be provided by photovoltaics at the end of the century was 0.1–1.0 quad. But the lower limit was based on competing with oil at $25 per barrel. By 1980, oil was selling for $40 a barrel. Although prices subsequently fell to the low thirties, further escalations in oil prices are probable.

The upper projection of 1.0 quad would be realizable "within the framework of traditional Federal intervention . . . with a set of comprehensive and aggressive initiatives";[23] translated, this meant the relatively modest level of support would continue. The DPR projected 2.5 quads by the year 2000 as the "technical limit" for photovoltaics, although significant achievements in recent years indicate that this number may be appreciably exceeded.[24]

The net result of the DPR was to provide the lowest common denominator of the assorted participants rather than a challenge to the nation. Amory Lovins called the DPR "a largely vacuous and uncreative exercise." Congressman Robert Drinan termed the Carter administration's position "Hypocrisy . . . characterized by glowing rhetoric on solar energy, but little meaningful action." The wide publicity given the DPR left the public with the impression that photovoltaics has little to offer in this century.[25]

The DPR and the projections by the Council on Environmental Quality are of continuing interest because they are the only official estimates of the commercial potential of photovoltaics that have been made by government agencies. In its zeal to squelch photovoltaics—and all forms of solar energy—the Reagan administration took the position that government should have no active role in its commercialization but leave this aspect entirely to the private sector. (Interestingly, the Reagan people continued to request and get large appropriations for research and development on nuclear fission.)

A study, completed in January 1979 under the auspices of the American Physical Society (APS), concluded that the cost reductions required to make photovoltaics practical are indeed feasible. It examined only the central-station, utility-operated application, however, and found that "It is unlikely that photovoltaics will contribute more than about 1% of U.S. electrical energy by the end of the century." By recent projections, this would be somewhat less than 1 quad. At the time of the study, central-station use was thought to be farthest in the future; this is no longer the case, as utilities are already contracting to buy power from central plants built through third-party financing. The APS study no doubt reinforced the impression left by the DPR that photovoltaics is at best a long-term solar option.

A favorite technique in analytical studies of future market opportunities is to calculate the break-even price for photovoltaics—the price of equipment in dollars per Watt at which the cost of electricity from photovoltaics will be the same as the cost from the existing source with which it must compete. It is assumed that significant market growth will begin for that particular application at that point. In the real world, however, the market usually develops much sooner, when the price is still two or three times higher than is required for direct competition. Simple hand-held calculators that can be bought for a few dollars today were selling in the late 1960s for $200 because to a few people they were quite useful. A complex set of factors motivates early buyers. Sophisticated analytical constructs often contain such inconspicuous yet crucial oversimplications.

With widespread commercialization of photovoltaics yet to come, we should examine present and near-term growth rates and ask what factors, characteristics, and conditions are inherent in the technology that may deflect it from its present course—in effect, explore the problem in the negative.

The goal of 4 peak gigawatts total cumulative production by 1988, set by the Solar Photovoltaic Energy Research, Development, and Demonstration Act of 1978, represented a consensus of knowledgeable people in industry, government, and the research community at that time. The Act assumed a continuing doubling of production annually, which was a reasonable extension of the current trend.

Such a growth rate, when plotted on semilogarithmic paper, yields a straight line. Figure 6-1 depicts this in that portion of the upper curve extending to 1988. This growth pattern is typical of new industries or enterprises—even of living things, be it a plant or child. Since 1983, decreased government emphasis on solar energy, coupled with the oil glut and a recession (both of which eased the growth rate in electricity demand), have caused several years' slippage in the goals of the Photovoltaic Act. Still, the prospects for growth remain good. Let's review the factors that could limit the growth and see which ones—if any—pose a real threat.

Continued slack in the demand for electricity could limit the potential growth of photovoltaics. However, both an increasing population and the steady trend toward urbanization make electricity a preferred energy form, especially in densely populated areas. If adequate quantities can be provided, the use of electricity will continue to grow. Whereas today we are forced to curb the use of electricity, photovoltaics will make wider usage possible.

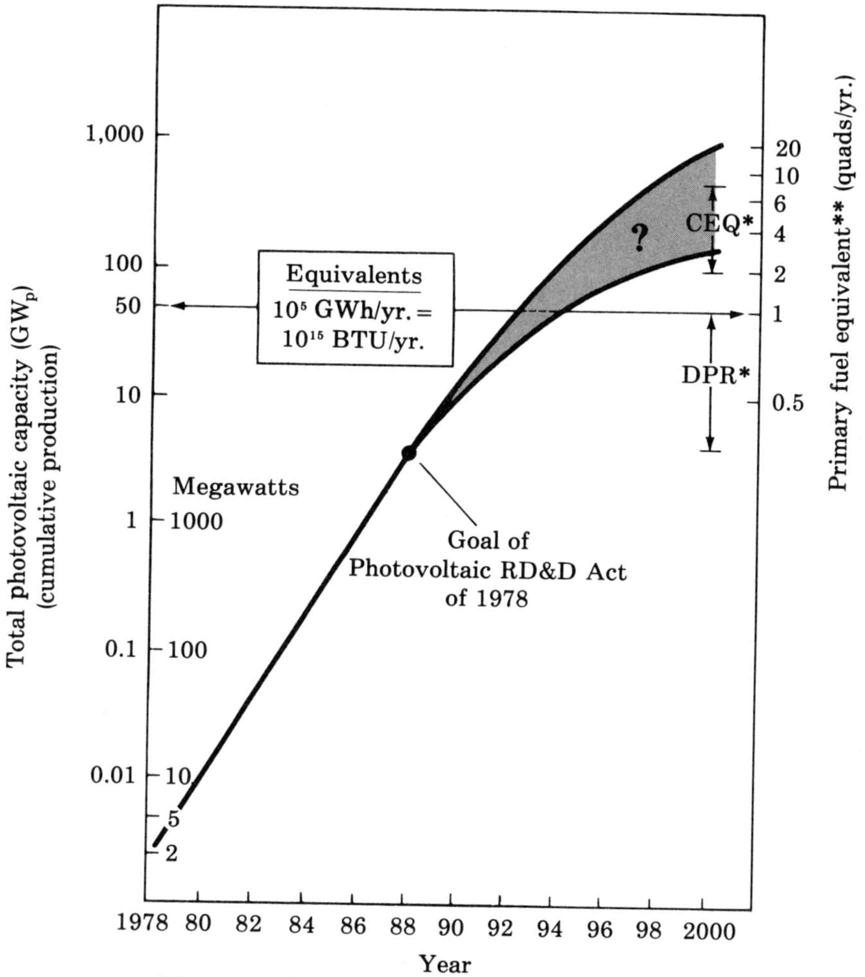

*Estimates of:
CEQ - Council on Environmental Quality
DPR - Domestic Policy Review

**Basis: 1 kWp photovoltaic array produces 5.48 kWh/ day, 2000 kWh/yr.
1 kWh = 10,000 BTU of primary energy.

Figure 6-1 Growth possibilities to the year 2000 for solar photovoltaic energy systems worldwide. It appears the goal of the Solar Photovoltaic Energy Research, Development, and Demonstration Act of 1978 to achieve cumulative U.S. production of 4 gigawatts by 1988 will slip six to eight years. Meanwhile, production in Europe and Japan is accelerating. Eventually widespread application of the technology seems assured. Estimates of the Council on Environmental Quality and the Domestic Policy Review are indicated at right. (Vertical scale is logarithmic.)

The growth of some technologies is inherently limited. But because the photovoltaic phenomenon occurs in an extraordinarily thin layer of material, and because a number of materials work quite well, it is a fertile field for exploitation. The materials are usually plentiful and relatively inexpensive. While balance-of-system components are not subject to similar dramatic improvements, mass production and design simplifications will permit continued decline in system costs. The long-term growth potential of photovoltaics is clearly not limited by technological constraints.

Many technologies founder because very large units are required to make them economically attractive. Rapid scale-up in size is notorious for causing development delays, high failure rates, and large cost overruns. In the solar field, the experimental Barstow "power tower" in California is a case in point;[26] delays and cost overruns nearly killed it in 1980. But photovoltaic systems are about equally effective in any size. They are modular; additional units can be added like strings of lights on a Christmas tree. The risks inherent in extrapolating to very large units will not be an impediment in the application of photovoltaic technology.

None of the factors discussed above pose any real threat to the growth of photovoltaics. The factors we will turn to next may be more problematic.

The necessary extensive manufacturing facilities will take time to construct and equip. While this will not be a bottleneck, the sheer numbers involved will serve to moderate growth rates in the 1990s.

Despite extensive automation, factories will require large numbers of skilled workers and trained supervisors. It will take time to develop this work force.

Retail sales, installation, and service organizations will have to be formed. Financing institutions will have to learn about photovoltaics. The gradual buildup of this supporting infrastructure will serve to temper the growth rate.

Institutional obstacles such as zoning regulations and construction codes that have not yet been adapted to the technology will require time for adjustment.

In the long term, storage will become a significant factor in dampening photovoltaics' growth rate. A few hours of storage capacity for a residential system can be provided at reasonable cost today. But storage in the form of batteries or flywheels, while entirely feasible, requires components of significant size and weight. More efficient and less expensive batteries, flywheels, and other storage systems can be developed, but in general, storage devices are not subject to miniaturization per unit of energy stored.

In many applications, such as water pumping and desalination, and in some industrial uses, no storage is needed. For others, storage is a convenience or a way of making a photovoltaic system more efficient. In a few applications, storage is critical if an alternate source of electricity is not available. Storage will never be a bottleneck to photovoltaics, but in the mid- to late 1990s, it will exert significant restraint on an otherwise explosive growth rate.

The ultimate limitation is market saturation. This will occur in stages as one market segment after another approaches full use of photovoltaics—say, 75 percent of total capacity. For example, photovoltaic houses are expected to become a major growth market in the 1990s, and by the turn of the century practically all new housing could routinely be fitted with photovoltaics. The rate of expansion in this field will then level off. Similar effects will occur with commercial buildings, perhaps following housing by a few years.

But even as one market segment is filled, another can open up. Photovoltaics may become the preferred energy source for processing synfuels from coal and oil shale. As things stand now, part of the coal must be burned to fuel the process. From one-half to one ton of coal must be consumed to convert another ton into other fuel forms. This inherent inefficiency is a major weakness of the synfuels effort. The process also intensifies the air pollution problem. Photovoltaics, probably as photovoltaic-thermal total energy systems, might render the synfuels venture considerably more attractive.

The international market—the developing countries—may reappear as the large market of the future. It will be years before many people can afford even the low-cost systems that will eventually be available.

In an uncertain world, however, such limiting factors—especially those responsible for the present relaxation—can be offset with startling suddeness. Political upheaval, international economic effects, the possible need for drastic action on environmental problems, even the public's appetite for new technologies can cause a surge in the use of photovoltaics. And the continued ingenuity of hundreds of scientists, engineers, and entrepreneurs plus the production and marketing drive of new companies will have their effect. Worldwide cumulative production by the end of the century may be in the tens and possibly hundreds of gigawatts.[27] In any event, the actual level achieved will depend heavily upon public understanding, interest, and support.

The question of market size is important. Perceptions of market growth significantly influence achievement of widespread use. Perceptions of market potential influence the level of government support and affect the commitment of funds by the private sector. Decisions hinge on estimates of future sales: The firms that make photovoltaic equipment, the distributors who sell it, and the banks that finance it will respond in accordance with their perceptions of market growth rates and the absolute size of the market. Such expectations also have a vast unseen effect on public enthusiasm and support for the technology.

President Kennedy said that space is the new ocean, and we would sail upon it. He set the goal of landing men on the moon and returning them safely to earth before the decade was out. The first moon landing was made eight years later. The challenge was backed by an unspoken yet profound expectation of fulfillment. Such prophecies tend to be self-fulfilling.

Photovoltaics will affect many aspects of society. This analysis has focused on the aspects that loom significant in the near term: supplies, jobs, health effects, the role of utlities, and broader issues of social equity. No crippling flaw is seen in this technology. On the contrary, photovoltaics will provide ample electric energy without pollution, freeing electric power usage from a conservation straitjacket caused by outmoded power sources and elaborate, costly distribution systems. Electric power will then be available to satisfy many needs, such as clean air in our cities and power for transportation. These will be critical factors in the years ahead.

NOTES

1. We base this projection on an optimistic plant utilization factor of 70 percent. The National Electric Reliability Council, Princeton, New Jersey, an industry group, projects 3.1 percent annual increase in electric consumption for the next ten years. (*New York Times,* 22 May 1984, pp. D-1, 4). Most authorities, including the Department of Energy, still predict at least a doubling of electricity use in the United States by the year 2000. Chauncey Starr, former president of the Electric Power Research Institute, says that by 2000, demand will be 2–2½ times the 1980 level. He admits to a large shortfall in supply. "Choosing Our Energy Future," *EPRI Journal* 5 (September 1980), pp. 6–11.
2. Raw and bulk materials requirements for large-scale use of photovoltaics have been extensively studied by Battelle Pacific Northwest Laboratories. See N. E. Carter, et al., *Some Potential Material Supply Constraints in the Deployment of Photovoltaic Solar Electric Systems,* PNL-2971 (Richland, Wash.: Battelle Pacific Northwest Laboratories [1978]), pp. 1–46; and R. L. Watts, et al., *The Evaluation of Critical Materials for Five Advanced Design Photovoltaic Cells with an Assessment of Indium and Gallium,* PNL-3319 (Richland, Wash.: Battelle Pacific Northwest Laboratories [1980]), pp. 1–99.
3. U.S. Department of Energy (ERDA). *Solar Program Assessment: Environmental Factors Photovoltaics,* ERDA 77-47/3 (Washington, D.C.: Government Printing Office, 1977); and *Environmental Development Plan for Photovoltaics,* DOE/EDP 031 (Springfield, Va.: National Technical Information Service, 1979).
4. Council on Environmental Quality. *Solar Energy—Progress and Promise* (Washington, D.C.: Government Printing Office, 1978), pp. 2–3.
5. "Employment Impact of the Solar Transition," by Leonard S. Rodberg, Joint Economic Committee, U.S. Congress, 1979; Meg Schacter, "The Job Creation Potential of Solar Energy," Second Open Workshop on Solar Technology, Solar Energy Research Institute (Washington, D.C., October 1979).

6. National Science Foundation—Stanford Workshop on Net Energy Analysis. Institute for Energy Studies, Stanford University, and TRW System Group (Palo Alto, Calif., August 1975.)
7. D. W. Fraley, C. L. McDonald, and N. E. Carter, "A Review of Issues and Applications of Net Energy Analysis." (Richland, Wash.: Pacific Northwest Laboratories, 26 September 1977).
8. In computing energy payback time, allowance is usually made for the energy consumed in producing the raw materials used.
9. ABC, "Issues and Answers," (29 June 1980). Interview of G. William Miller. Transcript from Tyler Business Services, Washington, D.C., 1980.
10. Carrol L. Wilson, "Coal: Bridge to the Future," Report of the World Coal Study (Cambridge, Mass.: Massachusetts Institute of Technology, 1980).
11. The Brandt Commission was an independent international group headed by Willy Brandt, former chancellor of West Germany and holder of the Nobel Peace Prize.
12. Private communication with Royal Dutch Shell.
13. Private communication.
14. "Hayden and Lovins Debate Institution Roles." *Solar Engineering* (October 1979), p. 24.
15. At a conference of energy specialists at Stanford University in June 1980; as reported by E. Marshall, "Planning for an Oil Cutoff." *Science* 209 (11 July 1980), p. 246.
16. Marshall, "Planning for an Oil Cutoff," *Science*.
17. BDM Corp. and Solarex Corp., *Photovoltaic Power Systems: Market Identification and Analysis—Final Report,* Vol. 1: *Executive Summary and Main Report,* prepared for U.S. Department of Energy, Office of Energy Technology, Division of Solar Energy, contract EG-77-C-01-2533 (McLean, Va.: BDM Corporation and Solarex Corporation).
18. Dennis Costello, David Posner, Dennis Schiffel, James Doane, and Charles Bishop, *Photovoltaic Venture Analysis: Final Report,* 3 Vols., SERI/TR-52-040 (Golden, Colo.: Solar Energy Research Institute, 1978).
19. *Solar Energy—Progress and Promise,* Council on Environmental Quality (Washington, D.C.: Government Printing Office, 1978), p. 6.
20. *Domestic Policy Review of Solar Energy.* Final Report: Research, Design and Development Panel, TID-28837, October 1978; and *Domestic Policy Review of Solar Energy,* TID-28834 (Springfield, Va.: National Technical Information Service, 1979).
21. President's Message on Solar Energy, The White House (Washington, D.C., 20 June 1979).
22. See "Fact Sheet—The President's Message on Solar Energy," Office of the White House Press Secretary (20 June 1979).
23. *Domestic Policy Review of Solar Energy,* p. 47.
24. The uncertainties in the projections were acknowledged: "It is not possible to forecast total energy demand or the future use of specific energy sources beyond the near-term with any certainty. A number of unpredictable factors such as the course of energy prices, the availability of competing fuels, future environmental standards, public attitudes, and the effects of government activity will affect the pattern of future energy use." *Domestic Policy Review of Solar Energy,* p. 47.
25. For an excellent short review of recent solar policy, see Steven Ferrey, "Solar Eclipse: Our Bungled Energy Policy," *Saturday Review* (3 March 1979), pp. 24–26. Ferrey served on the Domestic Policy Review.
26. Electricity is produced indirectly from solar energy. A field of mirrors focuses sunlight on a central boiler located atop a tower some 200 feet high, producing steam, which in turn drives a turboelectric generator at the base of the tower.
27. Equal installed operating capacity should lag this by not more than one year, as the bulk of production is near the end of the scale.

FURTHER READING

A New Prosperity—Building a Sustainable Energy Future. Solar Energy Research Institute. Andover, Mass.: Brick House, 1981.

Brooks, Paul. *Speaking for Nature.* New York: Houghton Mifflin, 1980.

Carter, Luther J. "Environmentalists Seek New Strategies." *Science* 208, May 2, 1980, pp. 477–478.

Characterization and Assessment of Potential European and Japanese Competition in Photovoltaics. U.S. Department of Energy/SERI/Science Applications, Inc. SERI/TR-8251-1. Golden, Colo.: Solar Energy Research Institute, 1979.

Courrier, Kathleen, ed. *Life After '80: Environmental Choices We Can Live With.* Andover, Mass.: Brick House, 1980.

D'Alessandro, Bill. "Photovoltaics: Milestones Passed, Miles Ahead." *Solar Age,* December 1979, pp. 12–14.

DeMeo, E. A., and Taylor, R. W. "Solar Photovoltaic Power Systems: An Electric Utility R&D Perspective." *Science,* April 20, 1984, pp. 245–251.

Environmental Development Plan for Photovoltaics. U.S. Department of Energy. DOE/EDP-0031. Springfield, Va.: National Technical Information Service, 1979.

Hayes, Denis. *Rays of Hope.* New York: W. W. Norton, 1977.

Hoyle, Fred. *Energy or Extrication?* London: Heinemann, 1977.

Ihara, Randal H., and Reid, Herbert G. "Technology, Democracy, and Solar Energy: The Moral Ambiguity of Solar Energy." *Proceedings, Helios: From Myth to Solar Energy,* 16–18 March 1978. Compiled by M. E. Grenander. Albany, N.Y.: State University of New York, Institute for Humanistic Studies, 1978.

International Photovoltaic Program Plan. 2 vols. Washington, D.C.: U.S. Department of Energy, 1979. Vol. I: *Report to the Congress in Accordance with the Solar Photovoltaic Energy Research, Development and Demonstration Act of 1978.* Vol. II: *Appendix.*

[Jackson, Bruce]. "Utilities: A Growing Solar Program." *EPRI Journal,* December 4, 1979, pp. 29–30.

Kidder, Tracy. "The Future of the Photovoltaic Cell." *Atlantic,* May 1980, pp. 68–76.

Kroemer, Sandy F. *Solar Law.* New York: McGraw-Hill, 1978.

Lönnroth, M.: Johannson, T. B.; and Steen, P. "Sweden Beyond Oil: Nuclear Commitments and Solar Options." *Science* 208, May 9, 1980, pp. 557–563.

Lovins, Amory, B., and Lovins, I. Hunter. *Brittle Power.* Andover, Mass.: Brick House, 1982.

Marvin, H. H. "Solar Energy in Our Future." *Proceedings, Helios: From Myth to Solar Energy,* 16–18 March 1978. Vol. 2. Compiled by M. E. Grenander. Albany, N.Y.: State University of New York, Institute for Humanistic Studies, 1978, pp. 304–314.

Morgan, Robert P. "UNIDO and Appropriate Industrial Technology." Editorial. *Science* 203, March 2, 1979, p. 835.

"Net Energy in Photovoltaic Array Manufacture." Rockville, Md.: Solarex Corp., 1978.

Novick, Sheldon. *The Electric War.* San Francisco: Sierra Club Books, 1976.

Partain, L. D. "Six Kilowatt Residential Photovoltaic Power Systems Study: Design, Performance, Economics, Market Potential." Livermore, Calif.: Lawrence Livermore National Laboratory, 1980.

"Photovoltaics in the Pacific Islands." DOE/CE/15095-5. Washington, D.C.: U.S. Department of Energy, September 1983.

Public Opinion on Environmental Issues. Council on Environmental Quality, et al. Washington, D.C.: Government Printing Office, 1980.

Reducing U.S. Oil Vulnerability—Energy Policy for the 1980's. U.S. Department of Energy. DOE/PE-0021. Washington, D.C.: Government Printing Office, 1980.

Solar Photovoltaic Energy Conversion. New York: American Physical Society, 1979.

Walton, A. L., and Warren, E. H., Jr. *The Solar Alternative—An Economic Perspective.* Englewood Cliffs, N.J.: Prentice-Hall, 1982.

Weiss, Charles, Jr. "Mobilizing Technology for Developing Countries." *Science* 203, March 16, 1979, pp. 1083–1989.

CHAPTER 7
INTERNATIONAL DEVELOPMENTS

Technology knows no national boundaries. Prior to the first oil embargo in 1973, steps were taken in both Europe and the United States to adapt photovoltaics for use on the earth.[1] The European Economic Community (EEC)[2] has diligently promoted photovoltaics in a program that closely parallels the U.S. effort. More recently, Japan has joined the parade, and several developing countries are taking first steps. A giant endeavor of international dimensions is being launched among the industrial and industrializing nations.

Conferences of the "photovoltaic specialists" in the United States, held at regular intervals for more than 20 years under the auspices of the Institute of Electrical and Electronic Engineers, are regularly attended by large delegations from Europe and Japan as well as scientists and engineers from other countries. A healthy rapport has developed among the participants, transcending national boundaries. Comparable technical meetings are convened periodically in Europe by the Commission of the European Communities, executive agency of the EEC, and U.S. personnel regularly attend and frequently present papers at these meetings. The Japanese Institute of Electrical and Electronic Engineers holds meetings on photovoltaics, to which persons from other nations are invited. A growing sense of competitiveness among corporate interests continues to be overshadowed by a strong international commitment to the technology and a determination to accelerate the development of photovoltaics in the larger interest of all.

Many companies have formed affiliates or opened subsidiaries across national boundaries. For example, Pilkington (U.K.) has invested in Solec International (U.S.), and Royal Dutch Shell is participating with Motorola to develop the ribbon-to-ribbon process. Meanwhile, Solarex (U.S.) has affiliations with companies in England, Belgium, the Netherlands, France, and Italy.

As major oil companies have entered the arena, the international structure of photovoltaics is being shaped as much by their corporate interests as by more commonplace national influences. The oil companies have provided risk capital for photovoltaics—an estimated $85 million in 1984—which has partially offset recent reductions in U.S. government support for photovoltaics. When these companies engage in marketing activities, they do it worldwide. In conducting a market survey of the use of diesel-electric generators—for which photovoltaics is increasingly an economic substitute—Shell Oil was able to call upon its dealers throughout the world for data. When large producers of photovoltaic equipment cut prices (an action most governments applaud), small photovoltaics producers have sometimes been squeezed out (a consequence most governments deplore).

It is fortunate that the oil companies are supporting photovoltaics, since venture capital is the limiting factor now in the spread of the technology. But it is most unfortunate that worthy entrepreneurial talent may be crowded out. Here is where a continued broad base of government support is sorely needed.

EUROPE

The photovoltaics movement in Europe is spearheaded by the Commission of the European Communities, which has been stimulating its development since 1975. In 1979, the Commission spearheaded a second four-year effort with 15.9 million European Currency Units (ECU), one-third of the total solar budget, earmarked for photovoltaics. (One ECU is approximately equal to 75 cents.) Since grants and contracts are usually on a matching-fund basis of 50 percent or more, actual commitment is much larger. Of the 15.9 million ECU, 10 million is being used for some 15 test and demonstration pilot projects. The remainder supports a variety of research and development activities. The third four-year plan, 1984–88, has equivalent funds, with less emphasis on demonstrations and more on export assistance.

The main emphasis is on systems development, including the fabrication and testing of prototypes suitable for commercialization. (This is in marked contrast to the United States, where the Reagan administration proclaimed systems development to be the domain of private industry and systematically dismantled the formerly balanced federal program; federal funds are now being restricted to research of a long-range nature.)

The Europeans are especially sensitive to their long-term energy needs. In order to maintain and improve living standards, Europe must operate its industrial base to produce exports; this results in high energy demand. For modest economic growth, Europe needs a minimum 2–3 percent annual

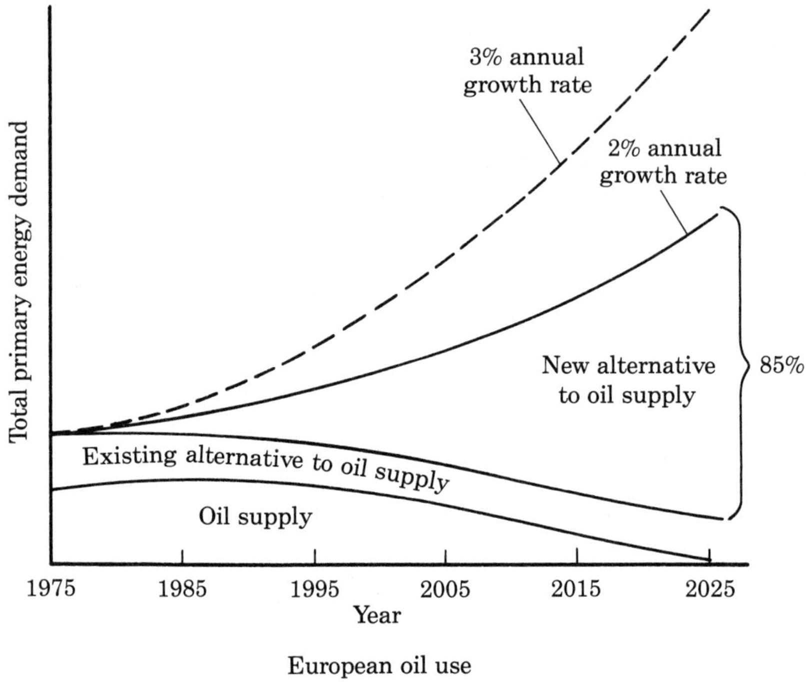

Figure 7-1 If total oil consumed in Europe declines as shown, a significant portion of replacement plus growth must come from renewables. (From *Photovoltaic Power for Europe* by Starr and Palz, with modification)

increase in energy supply. While coal reserves are adequate, costly deep mining is often required. Nuclear-electric expansion is restrained by limited uranium resources, plant-siting problems, and related social concerns. By traditional standards, oil prices are still outrageously high. Europe cannot look to imported oil for the increased energy needed; rather, predictions indicate a decline in oil usage (Fig. 7-1).

Under these circumstances, the EEC is actively promoting the development of renewable energy sources, including photovoltaics. If, as Michael Starr and Wolfgang Palz suggest,[3] only 10 percent of total primary energy supply would come from renewables by the year 2025, and 15 percent of that is provided by photovoltaics, 60 gigawatts of photovoltaics would be needed. Projecting from a 1-megawatt base of photovoltaic systems in Europe, they calculate the required rate of increase to be 30 percent per year, "a high but not totally impossible goal." Their calculation, which is based on present photovoltaic conversion efficiencies, appears to be quite conservative.

Most of Europe is well served by electrical grid. The isolated spots that still exist are sparsely populated mountainous regions and small islands, which rely largely on privately owned diesel generators. The central power plants are usually state owned, and grid supplies are highly controlled and quite reliable. Voltage, frequency, harmonic distortion, and power factors are kept within close tolerances, and safety standards are high. European authorities are therefore understandably hesitant to incorporate large numbers of dispersed generators in the form of privately owned photovoltaic, wind, or other small power sources into the main electrical grids. Moreover, in some countries privately generated electricity is prohibited by law from being sold either to the utility or to other consumers. The EEC is looking to the elimination of such regulations and to the alleviation of other technical and administrative constraints that hamper small power generators.

As in the United States, the price that utilities pay private producers for excess power is crucial to the growth of photovoltaics. But in Europe every nation is free to set its own buyback rates. (Throughout the United States, the law is that the buyback rate must at least equal the "avoided cost" to the utility.) In Denmark, the utility recently agreed to pay 30 percent of its normal sale price to a cogenerator for power returned to the utility. In the United Kingdom, buyback rates are based on the avoided cost to the utility, but each case is subject to separate negotiation based on the specific circumstances.

Utilities are already pondering the effects of tying in large numbers of small independent producers. The Italian utility ENEL (Ente Nazionale per L'Energie Elettrica) estimates that 1 megawatt of solar capacity would save it 0.3 megawatt of conventional generating capacity, at least for the first several percent of penetration by solar.

Table 7-1 Solar Resource Comparison (kWh per sq meter per year; fixed tilted array)

Europe		*North America*	
London	1,049	Northern U.S.	1,650
Copenhagen	1,182	Southeastern U.S.	1,890
Dublin	1,233	Southwestern U.S.	2,310
Paris	1,240	Southern Canada	1,430
Bonn	1,107	Mexico	1,800
Madrid	1,807		
Rome	1,852		
Athens	1,927		

All these factors point to slower introduction of photovoltaics in Europe than in the United States. In addition, Europe, particularly the northern part, does not have nearly the solar resource of the United States. (Some values are given in Table 7-1 for comparison.) For this reason also, the market for photovoltaics in Europe might be expected to lag that in the United States by several years.

Nevertheless, the market for residential grid-connected systems in Europe is expected to grow to 15 megawatts per year by 1995 and to be matched by an additional 15 megawatts in the commercial-institutional-industrial sector. Both areas would continue exponential growth thereafter. The use of photovoltaics by the utilities themselves (central station concept) remains a matter of speculation, depending heavily on future national dictates and on physical factors such as the availability of land. Wolfgang Palz, who heads research and development on solar energy in the EEC, says: "Large photovoltaic central power plants in Europe and in Northern Africa may well have a larger potential to meet a significant part of Europe's energy needs than any other of the new energies, including fusion."[4]

Sale of stand-alone systems will be relatively uninhibited by comparison and will continue to grow in popularity. Small units (10–500 peak Watts) are expected to provide a market of 12 peak megawatts per year by 1995 and level there, while the demand for larger units (to 150 peak kilowatts, mainly replacing diesels) may reach 50 peak megawatts per year by 1995, with continued growth expected.

The total European market is thus predicted to reach well over 100 peak megawatts per year by 1995, growing exponentially to the 1,000 peak megawatt level by 2000. Palz and Starr predict that the total annual value of photovoltaic equipment produced in Europe by 2000 would be 5–10 billion ECU. They add: "By the year 2025, the total installed capacity of photovoltaic systems in Europe could exceed 200 peak gigawatts, generating about 10 percent of total electricity production."[5]

In addition to supporting contracted research, the EEC also sponsored 15 "pilot projects," which are active demonstrations of intermediate-size systems (30–300 peak kilowatts), similar to the U.S. PRDA projects. In addition, three smaller projects of residential size were constructed with EEC support (near Adrano, Italy, at the site of the Eurelios solar thermal experiment). All were in operation at the end of 1983. The pilot projects are physically located in eight of the ten EEC countries, plus one in French Guyana. Their aggregate power exceeds 1 peak megawatt.

The applications chosen for the European projects reflect a strong interest in the export potential of photovoltaics. Four provide power to isolated villages. Seven are located on islands. Two power water desalination—

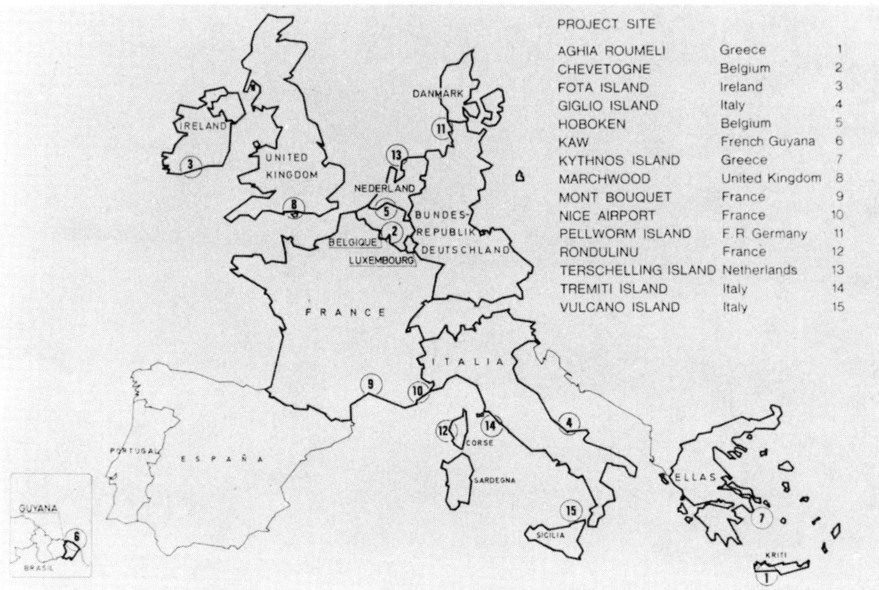

Figure 7-2 Fourteen photovoltaic demonstration pilot projects are in operation in eight European countries and one in French Guyana under the auspices of the Commission of the European Communities. (Courtesy: Commission of the European Communities)

disinfection systems. In contrast to the U.S. program, the EEC effort is oriented more toward stand-alone uses than grid-connected applications.

Total cost of the pilot program was about $30 million, with the Commission bearing one-third of the expense and the remainder being contributed by the respective host countries or organizations.

The Commission also operates a Joint Research Centre at Ispra, Italy. Finally, the Commission supports research and development on solar cells and modules through contracts directly with industry and research institutions, generally on a 50–50 cost-sharing basis. There are 50–60 active research and development contracts.

The Commission's total budget for photovoltaics for the years 1979–83 was only about $17 million, modest compared to the U.S. DOE effort for the same period, which exceeded $400 million. But the European industry since 1980 has become "clearly motivated and strong," says Palz. In addition, several individual countries have their own national programs. For example, Italy, through its energy research and development agency, ENEL, had a budget of $70 million for photovoltaics in 1982–84. A laboratory has been

Figure 7-3 Largest pilot project of the European Economic Community is this 300-kilowatt photovoltaic installation, which provides power at Pellworm Island, Federal Republic of Germany, a popular North Sea resort area. Cattle graze near the array. (Courtesy: Commission of the European Communities)

built in Naples, and a number of research and development projects are being supported elsewhere. The largest is the planned Delphos project, a 1.15 peak megawatt unit, which will supply a village of 800 inhabitants in Foggia (Apulia), 20 kilometers south of Rome. Delphos is to be funded 50-50 with the national electric utility, ENEL.

Other countries have programs of more modest size. Total governmental support including EEC efforts for photovoltaics in Europe is currently estimated to be about equal to that of the United States.

European authorities are fully cognizant of the fact that, lacking political unity, they face special challenges in reaping the benefits of this new technology. The excellent progress they are making is a tribute to their capacity to work across national boundaries for the common good.

JAPAN

The development of photovoltaics in Japan has been slower than in the United States. Japanese interest flowed naturally from that country's work with silicon in the semiconductor field, and Japan found extensive application in products of that field such as a method of powering small handheld calculators and other portable consumer products like radios, wristwatches, and toys, either with or without batteries. (Items that must operate continuously employ batteries recharged by photovoltaic cells.)

Because such articles require very little power, the Japanese were motivated to exploit the least expensive form of photovoltaic device, the amorphous silicon cell. The low light conversion efficiency and shorter lifetime of amorphous cells were no impediment here.

The Japanese are now the leading producers of amorphous silicon cells and are ahead of the United States in marketing and testing amorphous silicon products. The United States is considered about on a par with Japan in amorphous silicon cell research and general understanding of amorphous silicon technology.

More recently, the Japanese have proceeded to exploit the semicrystalline and crystalline technologies for use both at home and for export trade.

Government support for photovoltaics is centered at cabinet level in the Ministry of International Trade and Industry (MITI), where it is under the Agency of Industrial Science and Technology. In 1974, the "Sunshine Project" was formed as the focus for all energy research and development other than nuclear, and photovoltaic research is now funded through it. Recent support for photovoltaics increased steadily from $11.9 million in 1980 to $35 million in 1984.[6] Part of direct labor costs are funded by national laboratories and participating industrial firms, and it is estimated that the total commitment to photovoltaics in 1984 was about $80 million.

The primary goal now is to assure the availability of price-competitive photovoltaic equipment for general grid-connected usage by the latter 1980s. A stated objective of the Sunshine Project is to deploy 3,000 peak megawatts (3 peak gigawatts) of photovoltaics by 1990, the equivalent of powering 3 percent of all residences plus 7 percent of large buildings with photovoltaics. The Japanese estimate this would save some 82,000 barrels of oil a year. This would be expanded to 7,500 peak megawatts by 1995. A long-term goal is to derive 15 percent of all electricity from roof-mounted photovoltaic systems.

Research is performed largely by universities and special institutions. Some six universities are involved. To accelerate technology development and applications of new energy sources, the New Energy Development Organization (NEDO) was set up in 1974 within MITI. In 1984, it had contracts with some 20 companies.

The government–industry relationship on photovoltaics is handled somewhat differently in Japan than in the United States and Europe, where contracts are let competitively with industry for specific technology-development activities. In Japan, individual companies, universities, or laboratories are assigned overall responsibility for specific areas such as refining low-cost silicon, cell formation, module encapsulation, or ribbon development. The lead organization is then expected to share its findings so that commercial processes can be used up and down the line for greater efficiency and increased economic leverage.

It is estimated that about three-fourths of Project Sunshine funds go to technology development of modules and systems development. The remainder is committed to research and development on thin films, including amorphous silicon.

Sharp, specializing in crystalline silicon cells, continues to be the leading photovoltaics producer in Japan, followed closely by Sanyo in amorphous silicon. Japan Solar Energy Company is developing EFG ribbon. Sharp recently added an amorphous silicon capability through a joint venture with a U.S. firm, Energy Conversion Devices.

The Japanese are moving presently toward larger portable power units, which can be marketed as discrete packages for any appropriate use. Kyocera (Kyoto Ceramic) recently introduced a portable power module consisting of a nickel–cadmium battery with integral photovoltaic recharger. Nearly 300,000 units were sold in 1984. A market survey by Sanyo of the consumer products area forecasts that by 1990, in this one field alone, 50 megawatts per year of photovoltaics may be used.

Figure 7-4 Fifty-kilowatt array on the roof of a new dairy building on Fota Island, Ireland, provides power for milking machines, chillers, and other machinery. More power is required in summer than winter, tying this application to availability of sunlight. The project is monitored by University College, Cork County. (Courtesy: Commission of the European Communities)

The cost of purified silicon in Japan remains high. Excessive energy costs are a contributing factor. A pilot plant, soon to be completed, is looking to increased production of pure silicon at reduced cost.

As yet, the Japanese are not doing much work on concentrators, nor are the Europeans.

Japan's output of photovoltaic devices in 1984 was about 2 megawatts of crystalline cells and about 6 megawatts of amorphous units. By 1990, annual production of photovoltaics is expected to be about equal to that of the United States.

OTHER COUNTRIES

As photovoltaic technology continues to surge forward, people everywhere will become deeply interested. The developing countries constitute a vast potential market. Until recently, it was thought that they would make up the

Figure 7-5 In Kaw, French Guyana, a small village of 70 inhabitants obtains power for lighting, refrigeration, and communications from a 35-peak-kilowatt photovoltaic system. It is considered competitive with diesel generators at this location due to rising operation and maintenance costs of the latter. The array is positioned horizontally in this equatorial location. (Courtesy: Commission of the European Communities)

earliest market of large size. This assessment is now in question because of recent moves by several U.S. utilities, or third-party owners in deals with utilities, to open multimegawatt central station plants in the United States. The outcome will depend partly on the relative cost of photovoltaics compared to diesel-electric sources, partly on access to financing by the developing countries, and partly on the range of factors affecting power mix options in the United States.

Rather than depend upon others for access to this new medium for tapping the sun's energy, developing countries will increasingly seek to supply their own growing demand for photovoltaic equipment. They can get into the business at any of three levels, depending upon their degree of industrialization, and develop their own capability to install and maintain photovoltaic systems. They can buy complete systems. They can procure components of systems abroad and establish assembly plants, thus obtaining a partial production capability. They can set up plants to produce cells and modules, which involves high technology.

The obverse of the problem facing the developing countries is vividly depicted in the experience of the Spire Corporation of Bedford, Massachusetts. Founded by Roger Little in 1969 to do consulting and energy research for the aerospace industry, Spire eventually found its way into the solar cell business. By 1973, it had developed new manufacturing techniques for photovoltaics. The small, high-technology firm faced a dilemma. Says Little, "We could have tried to manufacture solar cells, but we recognized that the competition was just too awesome. We could have gone into systems and offered solar-powered water pumps internationally, but we didn't have a large marketing staff or experience. We chose the equipment route because that's what we knew best, and we could do it with less capital."[7] Now Spire is literally in the business of selling the technology, designing and setting up complete production facilities for photovoltaic cells, modules, and systems for anyone who wants to enter the business.

Spire has sold module assembly plants to Brazil and India. Pakistan and Tunisia reportedly are interested. But to Little, his biggest achievement to date is an agreement reached in 1983 with a company in Saudi Arabia, which has the license to manufacture photovoltaics in that country. (The Saudis already have in operation a large 350-peak-kilowatt system for village power.) Instead of a fee, Little reportedly settled for a royalty under the license. He says Egypt, India, and Pakistan are now interested. "We are the only company in the world," says Little, "that exists to put other people in the photovoltaic business. We want to be the McDonald's of solar cells. . . . What I'm responding to is . . . world trends. I'm not creating anything new."

Mexico is introducing solar power for village use in their Suncalli Project. Telefunken of Germany installed in Mexico nearly 200 kilowatts of photovoltaic systems for village power and trained local talent to operate them. Mexican officials have made it plain, however, that they intend to manufacture their own equipment as soon as possible.[8]

In Brazil, Heliodinamica, S.A., of Sao Paulo, is reportedly the first and only company in the Third World that has a complete manufacturing capability beginning with silicon and extending to finished photovoltaic systems. In 1981, Heliodinamica installed its first water pumping system in Rio Grande do Norte and is now selling to the government telecommunications company, which is replacing diesel generator systems with photovoltaics.

Dr. Lloyd O. Herwig of the U.S. Department of Energy states:

> In 1980, there were only about 5 module producers outside the U.S., who shared less than 15 percent of the world market. These companies

were located in Europe (largely France and West Germany) and Japan. There are now estimated to be more than 35 overseas companies producing photovoltaic modules or actively developing facilities for production in the near future.

In Europe, 13 companies are producing flat-plate modules, as follows:

United Kingdom	1
France	2
Germany	3
Belgium	2
Italy	4
Spain	1

Additional companies are fabricating systems from purchased modules.

In Japan, seven companies are producing flat-plate modules or systems, and five are making small photovoltaic power units for consumer products.

In addition, there are three companies in India, and one each in Brazil, Australia, and Malayasia (Singapore).

Concentrators are being made only in the United States (three companies).

When major research and development activities are included, an approximate breakdown of photovoltaic facilities by regions, worldwide, is as follows:[9]

Africa	1	North America	34
Australasia	34	Canada	(1)
India	(5)	Mexico	(2)
Japan	(21)	United States	(31)
Europe	44	South America	2

Because energy is basic to wealth, power, and independence, Third World countries will want to "bootstrap" themselves to energy independence by obtaining indigenous manufacturing capability for photovoltaics any way they can. Inevitably, this will pose difficult questions in the industrialized countries regarding the export of high technology. Representative Berkley Bedell (D-Iowa) remarked at a Congressional hearing, "I am afraid that other nations are going to run us out of the international photovoltaic market using our own technology. I am afraid that U.S. jobs and U.S. exports will be transferred overseas to our competitors."

As the United States strives to formulate appropriate industrial policy for the late 1980s, the jobs issue looms large. In the case of photovoltaics, U.S. industry will continue to grow. The principal question is how fast. In any

Figure 7-6 On the Island of Kythnos, Greece, a 100-peak-kilowatt photovoltaic unit supplies power to the local grid along with diesel and wind-powered generators. The photovoltaic-wind combination may be useful on wind-swept islands. (Courtesy: Commission of the European Communities)

event most of the new jobs will not be in manufacturing photovoltaic systems. Dr. Joseph Lindmayer, co-founder of Solarex Corporation, third largest manufacturer of photovoltaic cells in the world, says that most of the jobs will be in the infrastructure required to sell, install, and maintain photovoltaic systems for millions of homeowners and industrial, institutional, and commercial users. Today that infrastructure is virtually non-existent; it is a potential boom sector waiting to happen.

The problem for the United States is to maintain its present position of leadership across the whole field of photovoltaics. Within that framework, we can afford to help other nations establish indigenous capabilities matched to their technical capacity and other interests. The technology is so rich with opportunities—from research to manufacturing to applications—that pre-eminence must be maintained across the entire front. In the face of this, the Reagan administration in fiscal years 1981–84 consistently advocated further cuts in the federal effort—altogether it requested only $32.7 million for fiscal year 1984—asserting that everything except long-term, high-risk research should be left to the private sector. Meanwhile, the administration staunchly advocated further commercialization supports for the nuclear industry.

A leading newspaper editorialized:

> The Reagan administration's hopes and plans for solar energy diverge startlingly from those of the public—and from those of forward-looking foreign governments. In this country, public support for solar energy continues to be so strong that political consultants advise their clients to sponsor a solar bill as a sure way of winning constituent favor. But in the White House and the Department of Energy, it is unaccountably ignored, even sneered at.
>
> The president calls solar energy "exotic," by which he apparently means either that it is not widely applicable or that it is excessively sophisticated. Neither formulation is correct. France, Japan, West Germany and others are accelerating their solar research and are threatening to erase the large technological edge the United States enjoyed a few years ago.
>
> Solar energy will eventually make it on its own, but years or perhaps decades later than it could have.[10]

Fortunately, under the enlightened leadership of Paul Hodel and Pat Collins, the attitude toward solar energy at DOE later improved.

One of the beauties of photovoltaics is that it is growing far and wide. Demand for reliable photovoltaic equipment that is cost competitive with other energy sources will outstrip supply at least for the rest of this century. Peoples everywhere will want dependable photovoltaic systems at prices thev can afford. In the long run, the international marketing system will prevail, assuring that those companies most effective in supplying good equipment at reasonable prices will survive, irrespective of national location.

In all of this, the humanitarian aspect of the photovoltaic revolution needs to be kept in mind. So far, the United States has not been lacking in this regard. In a project managed for the U.S. Department of Energy by the NASA Lewis Research Center, the application of photovoltaic technology to meet special needs of developing countries has been under way for more than five years. Refrigerator-freezers for the storage of vaccines and medicines have been developed in conjunction with the Centers for Disease Control and the World Health Organization and are now being tested in 28 overseas locations under the auspices of the U.S. Agency for International Development. The Centers for Disease Control estimate 30,000 photovoltaic-powered refrigerators will be needed in its near-term program.

Portable photovoltaic power packages are being introduced for other medical applications such as sterilizers, dental equipment, and X-rays. Also through AID, photovoltaic equipment is being adapted to meet power needs in remote locations for education, communication, water purification and pumping, food processing, and lighting, as well as to promote the growth of cottage and rural industries.

All nations can participate eventually in the commercialization of photovoltaics, and all will benefit from its use. Joe Lindmayer, refugee physicist from Hungary who started Solarex in his basement, puts it simply: "It's apple pie. It's the flag. It's motherhood. You can't be against it. It's really that beautiful."[11]

NOTES

1. At the IEEE Photovoltaic Specialists Conference in Silver Spring, Maryland (1972), and at the UNESCO conference, "Solar Energy in the Service of Mankind," in Paris (July 1973). See W. Palz, "Photovoltaic Outlook From European Community's Viewpoint," 15th IEEE Photovoltaic Specialists Conference, Orlando, Florida (June 1981).
2. The European Economic Community (EEC) was formed as a common market in 1958 by Belgium, France, West Germany, Italy, Luxembourg and the Netherlands, and joined in 1973 by Denmark, Ireland, and the United Kingdom, and by Greece in 1981.
3. M. R. Starr and W. Palz. *Photovoltaic Power for Europe: An Assessment Study.* (Dordrecht, Holland: Reidel, 1983), p. 88. They cite other studies.
4. Proceedings, 15th IEEE Photovoltaic Specialists Conference (Kissimmee, Fla., May 1981), p. 18.
5. Proceedings, 16th IEEE Photovoltaic Specialists Conference (San Diego, September 1982), p. 1210.
6. Based on exchange rate of 230 yen per dollar.
7. Tom Richman. "The Fight To Harness The Sun," *Inc.* magazine (May 1983), pp. 123–134. We are indebted to Mr. Richman for this and several other quotations and background material used in this section.
8. Private communication.
9. Ray Watts, Battelle Pacific Northwest Laboratory. Private communication.
10. *Washington Post,* 26 July 1982.
11. *Washington Post,* 29 May 1983, p. C-3.

FURTHER READING

Commission of the European Communities. "Crucial Choices for the Energy Transition: An Initial Evaluation of Some Energy R&D Strategies for the European Communities." Brussels: CEC (EUR 6610 EN), 1980.

Ferber, Robert R. "Status of Foreign Photovoltaic R&D, Manufacturing and Markets." Statement to Subcommittees of the House Committee on Science and Technology, June 3, 1982.

Herwig, Lloyd O. "International Activities in the Commercialization of Photovoltaic Power Systems." Proceedings of the Annual Meeting, American Solar Energy Society, 5:1045–1050, 1982.

Maycock, P.; Shimada, K.; Stirewalt, E. N.; and Hunt, V. Dan. *America Challenged: Photovoltaics in Japan*. Alexandria, Va.: Photovoltaic Energy Systems, Inc., 1982.

Ministry of International Trade and Industry. "On the Progress of the Sunshine Project." Tokyo: February 1981 (in Japanese).

Morrissey, S. "Inventing a Solar Energy Future While Dodging the Oil Companies" (Interview with Dr. Joseph Lindmayer). *Washington Post,* May 29, 1983, p. C-3.

Palz, W. "Overview of the European Community's Activities in Photovoltaics." Proceedings of the European Community Photovoltaic Conference, Stresa, Italy, pp. 3–8.

Palz, W., ed. *Photovoltaic Power Generation: Proceedings of the Final Design Review Meeting on European Community Photovoltaic Pilot Projects*. Dordrecht, Holland: Reidel, 1982.

Richman, Tom. "The Fight to Harness the Sun." *Inc.,* May 1983, pp. 123–134.

Schiller, D. and Larson, Ronal W. *Photovoltaic Applications for Remote Islands*. Denver: Black Hawk Associates, 1984.

Shimada, K. *Photovoltaic Research and Development in Japan*. Springfield, Va.: National Technical Information Service, 1983.

Starr, Michael R., and Palz, W. *Photovoltaic Power for Europe—An Assessment Study*. Dordrecht, Holland: Reidel, 1983.

CHAPTER 8
PHOTOVOLTAICS AND THE ELECTRIC AUTOMOBILE

The potential of photovoltaics is clear. The issue is not whether we as a nation will be able to obtain electricity directly from sunlight at prices that will be acceptable for everyday use. With continued modest support by the government, the break-even point at which photovoltaics becomes economic for all uses will, as we have suggested, probably be reached by the late 1980s. The entrance of photovoltaics into the energy market will be gradual at first and then accelerate; by 1995 a fundamental restructuring of the market will be under way.

The serious questions to be faced now are whether to make this happen sooner and how to focus this new technology to alleviate present and impending national energy problems.

As we turn away from an era of energy gluttony, more efficient use of energy offers great benefits. Efficient energy use involves changing both our equipment (buildings, their insulation, and their heating and cooling plants; the machinery of transportation; and industrial processes and equipment) and our habits. Both aspects deserve careful attention.[1] While we are still at our most wasteful, significant progress can be made rather easily; but in the long term, new achievement will become more laborious and expensive. The unit cost of conservation in terms of physical plant and social amenities (quality of life) can only go up. As one realist put it, "Conservation may be absolutely necessary as a tactic, but it is potentially disastrous as a strategy."[2] Conservation makes good sense, especially while we are still careless in our use of energy, but it must be supported by vigorous activity to assure that adequate energy is available from acceptable sources at reasonable prices.

A new administration came to Washington in 1981 dedicated to the production of more energy. "The government has acted on the principle that the way to deal with energy is to do away with it," wrote President Reagan's energy policy task force. "Instead of unleashing the resources of a wealthy nation, we have, in the name of saving energy for some unspecified future time, tucked energy away like a rare bottle of wine. . . . We reject the notion that the energy dilemma can be solved only by halting the use of energy. . . . Energy resources are valuable only if they are produced and consumed."[3] The Reagan administration then shifted emphasis away from conservation and toward production. Today conservation remains a desirable and necessary objective, but providing adequate quantities of energy from acceptable sources has become an equally important goal.

The potential contribution of competing energy technologies to the national energy supply affects the speed with which photovoltaics is commercialized. Research and exploratory development on each energy technology should continue at a level reasonably commensurate with its promise. Nuclear fusion work should continue until it is clear that further effort would be futile or until the end of the tunnel is in sight. Perhaps we should drill enough deep wells to remove all doubt about the potential of geo-pressurized gas. Maybe we should even bulldoze over a few more counties here and there until we have revealed fully the implications of massive strip mining and synfuel operations.

The appropriate test is whether technologies such as these make sense for the long run, and if not, whether they will divert us from other technologies that are better matched to foreseeable future conditions.

NATURAL GAS

The U.S. supply of natural gas is limited and has been greatly depleted over the decades. One hope has been that present sources would be greatly augmented by geo-pressurized gas—large amounts of methane dissolved in deep underground waters primarily along the Gulf Coast. Recent evidence, however, has raised serious questions about the amount that may be recoverable.[4] The latest drilling indicates that the methane is too dilute and contained in too many small reservoirs to obtain it in quantity, even if the price of natural gas should double. Informed opinion holds that the discovery of new gas will barely maintain production at present levels, and in any event, it is not an easy substitute for liquid fuels required in transportation. Synthetic natural gas similarly appears unpromising, at least for another decade.

COAL

In discussions of the nation's energy future, attention returns repeatedly to the major fossil energy resource, coal, which is in ample supply in this country.

The problems of mining and burning ever larger amounts of coal are becoming increasingly apparent.[5] In recent years about 150 fatalities have occurred annually in the mines, and long-term health hazards to workers and to the public at large are just now becoming fully evident.

Moreover, most of the coal must be hauled overland closer to the centers of population, where it is burned to create electricity or to provide direct heat. Transportation is becoming an increasing nuisance, especially for small towns in the Midwest through which dozens of trains pass daily.

If used for heat, coal must be burned at the point of use, increasing pollutants in populated areas. When coal is burned to make electricity, two-thirds of the heat energy is lost—which is why some have decried the steady trend toward expanded use of electricity.[6]

One major study of coal suggests that global usage may have to triple to meet world energy needs to the year 2000.[7] The Council on Environmental Quality says high coal usage could mean that 2,000 square miles of land would be strip mined in the rest of this century and that another 5,000 square miles (an area the size of Connecticut) would suffer subsidence. When the earth sinks as abandoned underground mines collapse, anything on the surface can be destroyed.[8] We can probably manage, albeit with considerable discomfort, the mining and hauling of the vast quantities involved, but the core issue is what to do about the products of combustion, which weigh more than the coal. For every ton of coal burned, nearly three tons of combustion products result (mainly carbon dioxide).[9]

Of most immediate concern are the oxides of sulfur and nitrogen and particulate matter.[10] The larger particulates are relatively easy to remove at the stack, but the microscopic particles are disseminated for hundreds of miles, with long-term implications for human health. Worse, sulfur and nitrogen oxides combine with water vapor in the air to form some of the strongest acids known. In recent years, high stacks have been built to reduce local pollution, but this has only transferred the problem elsewhere. Increased coal burning in the Midwest is now producing acid rain in the Northeast.[11] Since the mid-1970s, fish have disappeared from thousands of lakes in New England and Canada. Indeed, acid rain has now become a matter of serious contention in our relations with Canada, and the problem grows steadily worse.[12] Automobile finish is being damaged by acid rain in some areas of the Northeast, causing car manufacturers to provide owners with special instructions for protecting their cars. Acid rain from burning coal now threatens Northern Europe as well. It is possible to remove much of the acidic components of stack gases at considerable expense, but this raises the cost of electricity and is therefore economically and politically unattractive.

The major long-term issue that results from the burning of any fossil fuel—the buildup of carbon dioxide (CO_2) in the atmosphere—is unresolved.

Carbon dioxide is highly volatile, and there is no way it can be removed from the atmosphere on the massive scale involved. CO_2 in the atmosphere absorbs sunlight in several parts of the sun's spectrum. The earth must get rid of all the heat it receives from the sun, or it would get hotter and hotter. The earth reradiates much of the energy into space at the longer wavelengths. CO_2 lets in most of the sun's energy, which arrives in shorter wavelengths, but blocks some of the heat that the earth must get rid of. This phenomenon is commonly known as the greenhouse effect.

It has been estimated that a doubling of the carbon dioxide content of the atmosphere will lead to an average global temperature increase of 4.5°F.[13] Since the beginning of the industrial revolution, the CO_2 content of the atmosphere is estimated to have risen 13 percent. At present rates of fossil fuel consumption, it will increase another 12 percent in the next 20 years alone. The effects of a change of even one or two degrees on the weather could be severe. For example, it is conceivable that the polar ice caps could begin to melt, causing oceanic flooding worldwide. The warming would be greater at the poles than in the tropics. In the last two decades, the CO_2 content of the air at the South Pole has increased 5 percent.[14]

Another complication is that the oceans absorb carbon dioxide from the air, and the effect that significant changes in the CO_2 content of the oceans would have on marine ecosystems is unknown. Plant life does use carbon dioxide, but vast increases in forests would be necessary to counter significantly the buildup in the atmosphere. At the moment, economic expansion is causing the world's forests to be gobbled up at an alarming rate. Large areas of uncertainty exist regarding how much CO_2 the planet can tolerate.[15]

Heat from the sun literally churns the earth's atmosphere; it causes the winds to blow, and it evaporates ocean water, eventually releasing it as rain or snow. If the amount of heat trapped in the atmosphere increases, the churning effect may be expected to increase. Meteorologists are in broad agreement that major variations in weather patterns may ensue. For example, normally arid regions may receive more rainfall while other areas experience drought. It is interesting to note that the number of tornadoes—or at least the number reported—has been increasing for a number of years.

Table 8-1 Average Number of Tornadoes Reported in the United States

Interval	*Number per Year*
1946–55	247
1956–65	548
1966–75	669
1976–81	829

Meteorologists are not sure of the cause. It may be that a higher world population results in more frequent sightings and that better communications produce more reports. We do not suggest that there may be a cause-and-effect relationship between the increasing number of reported tornadoes and the rising CO_2 content of the atmosphere. Rather, we mention it to illustrate how difficult it may be to obtain incontrovertible proof that changes in the composition of the earth's atmosphere affect climate and by how much and in what ways.

If we wait for a large body of evidence to be amassed and widely accepted within the scientific community, it will be very late in the day to start doing something to remedy climatic problems. Several decades of prodigious action worldwide will be required to start bringing the carbon dioxide growth rate under control. We know enough now to insist that eventually—within a few decades—the burning of fossil fuels must be reduced. Is it too soon to begin planning how this shall be done?

The carbon *monoxide* level in the atmosphere also seems to be increasing, apparently at a rate of about 6 percent a year.[16] This may decrease the atmosphere's capacity to remove dozens of man-made trace gases.

Writing about air quality, Philip Abelson, editor of *Science,* has said:

> Humanity is in the process of conducting a great global experiment. If unpleasant side effects are encountered, they cannot be quickly reversed.[17]

SYNFUELS

For a few months in 1979, the magic word that was to signal an end to the fuel shortage was "synfuels." Synthetic fuels, or synfuels, is a generic term that includes gas and liquid fuels made from coal as well as oil made from oil shale. We would strip mine coal that lies near the surface in several western states, crunch it, and pressure cook it in closed retorts to produce an oily liquid akin to petroleum. This could be refined into gasoline and other liquid fuels. (In one fantastic scheme, the burning and distillation would be done *in situ* underground.)

Because carbon dioxide is released both in the production process and later when the fuel is burned, the total CO_2 produced and, of course, the energy released as heat is at least 1.4 times greater, per unit of useful energy, than in the direct use of coal. The only valid reason for tolerating this waste is the need for liquid fuels for transportation. Because of high costs and other problems, such as the use of water in states where water is scarce, the synfuels program was never expected to yield significant results before 1990.[18] Today, the program is essentially dead.

NUCLEAR FISSION

Since the dawn of the nuclear age in the 1940s, nuclear power has repeatedly been urged as a satisfactory alternative to fossil fuels for generating electricity. Beginning in the 1960s, the utilities began to turn seriously toward nuclear energy. By 1980, there were some 90 reactors in operation with 100 more under construction or planned. But nuclear technology has turned out to possess an Achilles heel few had anticipated, a shortcoming possibly as much sociological as physical but nonetheless real. People in substantial numbers have simply rebelled against nuclear reactors, largely because they fear the effects of potentially uncontrolled radiation. The repeated demonstrations at the Seabrook reactor in New Hampshire and the accident at Three Mile Island (TMI) in Pennsylvania are landmarks, symbolic of the issues involved.[19]

TMI in particular did more than produce a psychological shockwave. Pennsylvania authorities have refused to allow Metropolitan Edison to charge its customers for the costly accident through the mechanism of the "fuel adjustment clause," which heretofore permitted utilities to bill their customers for substitute power when a nuclear facility is down. But the taxpayers are footing most of the cost of the cleanup, which is really just commencing. The Department of Energy is budgeting tens of millions each year under the rubric of "learning from the experience."

Meanwhile, in New York, problems with the Shoreham nuclear power plant, which cost more than $4 billion, threaten to bankrupt Long Island Lighting Company (LILCO). The net result of such setbacks is that planning and construction of new nuclear plants has all but ceased. To give just one example, the Tennessee Valley Authority (TVA) has decided to abandon construction of four nuclear plants because costs have become prohibitive. The plant at Yellow Creek alone would cost $10 billion, more than the entire cost of the TVA network, they say. Even if a utility company wanted to erect a new nuclear plant, it would find investment money very hard to drum up.

The most critical problem facing the nuclear industry is how to dispose of nuclear wastes. At present, nuclear plants store their spent fuel rods on the premises in tanks of water. Older plants may have to discontinue operation unless either more storage is provided or chemical reprocessing plants are set up to dissolve the rods in acid, separate and concentrate the fission products, and consign them to some "permanent" burial ground. Because the products of nuclear fission will be radioactive for several hundred thousand years, scientists debate whether safe disposal is indeed possible. Meanwhile, the stockpile of nuclear debris continues to accumulate.

The nuclear breeder reactor, which would produce as much new fissionable material as it consumes, thus providing an open-ended energy source, continues in limbo for two reasons. In the hands of many nations, breeders, which produce plutonium, would be likely to contribute to the proliferation of nuclear weapons. In addition, plutonium itself is extraordinarily dangerous from a health standpoint. For these reasons, in addition to fear of nuclear reactors in general, civilized society is most reluctant to commit itself to long-term dependence on plutonium as its crucial energy lifeblood.[20]

NUCLEAR FUSION

There is always the possibility of a breakthrough in controlled nuclear fusion, the "clean" process in which heavy hydrogen (deuterium) is caused to react at extremely high temperature and pressure, forming helium and releasing energy in the process. Despite prodigious research, success in practical terms appears at least several decades away. There would still be problems with radioactive materials, as the equipment and containment housing become extremely radioactive. Moreover, fusion would entail a further commitment to the central station approach to electricity generation and distribution. The consensus is that nuclear fusion on a practical scale is unlikely until well into the next century. The Harvard study concluded that "In the United States there is simply no reasonable possibility for 'massive contributions' from nuclear power for at least the rest of the twentieth century."[21]

OIL

As everyone knows, the energy crisis of the 1970s was precipitated by the Arab oil embargo of 1973. The basic problem with oil is that the resource is finite and dwindling, and reserves are under concentrated ownership. Neither factor was suddenly erased by the so-called oil glut of the 1980s. In 1973, net imports of petroleum products were running 6 million barrels a day, and in the mid-1980s they were still between 4 and 5 million barrels a day. The United States' economy is being bled white to pay for the petroleum bought from abroad, to the tune of some $45 billion a year. The price is exorbitant.[22] The effect on our economy could not be more pronounced if we manufactured and gave away the equivalent worth of automobiles, houses, computers, groceries, and other goods every year. The annual trauma over a few billion dollars in foreign aid is a mockery by comparison. The disastrous inflation of recent years and the resultant economic dislocation were the fruits of the energy crisis.

The industrial world's continued heavy dependence on imported oil is of major international concern. Many of our allies—notably Japan and the NATO countries—are more dependent on foreign oil imports than we are. Moreover, rising oil prices keep the developing countries in turmoil since they are being priced out of the market for the oil they need for bare subsistence. Human concerns aside, we obtain raw materials from many of these countries, and they are a market for the products of our agriculture and technology.

It is widely conceded that a small perturbation in the patterns of oil supply and usage would produce another price panic. Indeed, the Arab oil embargo of 1973 was not so much politically inspired as the result of shock at the jump in American oil imports that year (as indicated in the second column of Table 8-2). Not until 1978 did we begin to turn the tide on imports. High oil prices and more efficient cars may keep matters under control for the moment, but at best these are stopgap measures against a trend that is long term. It is even being suggested that the private automobile may be headed toward a major decline.[23] And the staggering cost of imported oil is not the only issue.

> The important role that control of the oil fields played in the Shah's downfall meant that the oil fields in other countries would become, even more so than before, a prime target of political dissidents. A revival of Islamic fundamentalism, co-existing with radicalism, will ensure a rich supply of political dissidents in Iraq, the Gulf sheikdoms, even Saudi Arabia, the mother lode of oil on which the Western world depends for political moderation and increased supplies.[24]

Twice our nation has been hostage to the political vicissitudes of the OPEC world. We must not forget the extent to which we are still vulnerable. Experts tell us that the oil glut is at best temporary and at worst explosive at any time. Daniel Yergin of Harvard Business School emphasizes that the energy situation may again become quite dangerous: "The more likely flash point occurs when accidents interact with a market in which demand is rising—as was discovered in 1973 and 1979. Thus, the world enters the danger zone when economic activity and energy demands are on the upswing."[25] Similarly, the International Energy Agency has warned that the world could face a devastating oil crisis in the coming years, and the present glut will only serve to intensify the crunch.[26]

The Congressional Budget Office estimates that a conceivable supply cutoff might well reduce oil imports by 3½ million barrels a day and last for a full year. This would cause a $270 billion loss to the general economy, an increase of 2 percent in unemployment, and a 20 percent jump in inflation.[27]

Continued dependence on foreign oil works directly against our efforts to build a more secure and prosperous world. For this reason, every energy proposal should be tested against one overriding criterion: How much and how quickly will it reduce foreign oil imports?

Table 8-2 U.S. Crude Petroleum Usage (million barrels per year)*

Year	*Imports*	*Domestic Production*	*Total*	*Imports as Percentage of Total*
1970	483	3,517	4,000	12
1971	613	3,454	4,067	15
1972	811	3,455	4,266	19
1973	1,184	3,361	4,545	26
1974	1,269	3,203	4,472	28
1975	1,498	3,057	4,555	33
1976	1,930	2,968	4,898	39
1977	2,406	3,009	5,415	44
1978	2,261	3,178	5,439	42
1979	2,355	3,121	5,476	43
1980	1,890	3,138	5,028	38
1981	1,604	3,129	4,733	34
1982	1,273	3,157	4,430	29
1983	1,206	3,159	4,365	28

* *U.S. Statistical Abstracts,* 1979, p. 759, Table No. 1329; *Monthly Energy Review,* DOE/EIA, March 1984, p. 21.

In summary, oil and natural gas simply cannot be discovered fast enough to expand the day-to-day supply. In real terms, world production of both may start to decline within a few years. Artificial substitutes in meaningful quantity are not in the offing. Use of nuclear energy is unlikely to increase greatly, certainly not in time to ease current problems. Both the near-term and the long-range consequences of greatly increased coal production and combustion appear unacceptable. Through stringent conservation, the fuels we do use can be stretched farther, but the cost and difficulty of conservation increase with each new round.

TIME FOR A CHANGE

This is a sobering list of shortcomings, external costs and Faustian bargains. In the face of these realities, we should ask ourselves whether a major commitment to solar photovoltaics on an accelerated schedule is not warranted.

The United States and the world are facing a watershed decision on the nature of future energy sources. Either we shall commit ourselves to further expensive expansion along the same old fossil-nuclear track, or we must seek out a new track and begin shifting to it as soon as possible.

What has been lacking is some germinal idea, some reasoned strategy for removing ourselves from the present cul-de-sac.

We have been throwing ourselves at the energy problem, and in the process, working into a societal and psychological morass. It is little wonder, for none of the major energy sources being manipulated—oil, gas, coal, nuclear—will fit into the long-range future. Moreover, most of them are ill suited to the present as well. Our society continues to be buffeted as various competing groups tear at each other, trying to build a future from elements that simply do not align with long-term needs. It is imperative to look farther down the road.

The Harvard Business School study recognized the problem: "Broadly speaking," it said, "the nation has only two major alternatives for the rest of this century—to import more oil or to accelerate the development of conservation and solar energy. This is the nature of the choice to be made. Conservation and solar energy . . . are much to be preferred."[28]

The first order of business is freedom from dependence on supplies of petroleum from abroad. If, by some magic, petroleum use could be cut by 40 percent, this would eliminate all need for foreign oil for a few years at least. The problem with reducing oil use lies with the equipment that depends on it; the automobile alone now consumes about half again more than we import.

The fossil fueled automobiles used in this country need to be steadily and deliberately replaced with cars that don't use gasoline or other petroleum products for fuel.

ELECTRIC AUTOS: A MARRIAGE MADE IN HEAVEN

Until quite recently, any thought of widespread introduction of electric vehicles (EV) would have been amiss, not only because the state of the art was not sufficiently advanced, but because electric power was frequently in short supply at peak periods. In a situation of short supply, a massive turn to electric power for transportation was not credible. Fortunately, recent progress with photovoltaics now places the whole matter of electric supply in a new perspective.

Some of the circumstances that have significantly affected the outlook for electric automobiles are:

- A widespread realization that fossil fuels are finite, uncertain, and carry high external costs
- The technical feasibility of massive photovoltaic deployment
- The development of a new generation of storage batteries that will both lower the costs for electric vehicles and provide greater driving range

- The fact that the transportation sector of our economy alone uses more petroleum products than we import
- The fact that the total inventory of automobiles is replaced about every ten years

Photovoltaic cells would not be put on an electric auto, as the available area could not provide nearly enough electricity to propel the car. Instead the photovoltaic array would be attached to the car owner's home. In addition to meeting normal household requirements, most household photovoltaic systems of the future could produce enough electricity to power an automobile.

An electric auto gets 2½–3 miles from 1 kilowatt hour of electricity. Fifteen kilowatt hours, about the amount a typical American household uses each day (excluding space heating and cooling), would propel an electric auto about 40 miles. This is typical urban-suburban daily use of the automobile, amounting to 1,200 miles a month, or 14,000–15,000 miles a year. Thus, a photovoltaic array of about 325 square feet (30 square meters) would provide enough energy in a year to power an electric auto 15,000 miles. The annual fuel cost for an electric car would be about the same as the annual electric bill for a medium-size house (without heating or cooling).

If electricity from a utility costs seven cents a kilowatt hour (the present average for residences), then the price for fuel is about three cents a mile, compared to perhaps four cents a mile for a compact gasoline car that gets 30 miles to the gallon. Thus, the direct fuel cost for an electric car is less than that of a gasoline-powered auto of comparable size.

But electric cars operate on batteries, and batteries cost money. The state of the art in batteries is rapidly advancing, so that in the near future a set of batteries for a car will last for 15,000 miles before replacement, and cost perhaps $1,000. This amounts to about seven cents a mile for batteries, which would make the total cost of powering an all-electric auto about ten cents a mile. However, the price of gasoline (and of diesel fuel) will surely rise over the long run. Meanwhile, new kinds of batteries under development will use less expensive materials than lead and last longer (see Appendix 10). Since batteries can be made in any size and shape, subsequent replacement with cheaper, more efficient batteries presents no problem. In mass production, the electric car will be economically competitive with the gas-powered auto.

Electric vehicles are now in use in many places and have proven their reliability. In England, for example, many delivery trucks operate on electricity; Cloride and Lucas are now producing 100 electric vans a month. In the United States, there are now about 4,000 electric cars on the road.

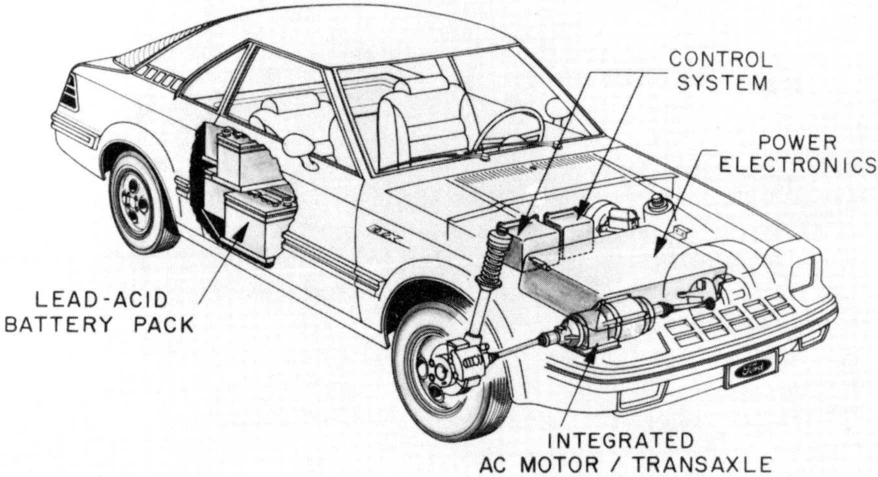

Figure 8-1 The first motor and transmission designed specifically for an electric car are shown as a single unit just behind the front bumper in this cutaway drawing. Housed together on one axle, the AC motor plus two-speed automatic transmission is only 9 inches in diameter, 20 inches long, and weighs only 90 pounds. An AC motor is preferred because it is smaller and lighter than a DC motor of equivalent power. (Courtesy: U.S. Dept. of Energy and the Ford Motor Company)

Close to 1,000 are produced annually, although production capacity is much larger than this. DOE has projected that some 8–9 million battery-powered electric vehicles may be in use in this country by the end of the century, given normal market growth.

While it is possible to convert a gasoline-powered auto to electric, it is not economically practical to do so. Tomorrow's electric auto will be an entirely new vehicle. A row of batteries will be placed low and on the centerline; this will lower the center of gravity for maximum stability. The cars will have specially designed, efficient motors. Electric vehicles traditionally have used commercially available electric motors. This situation is changing. Ford Motor Company with General Electric as a sub-contractor, and with DOE support, has developed an AC motor plus transmission all on one axle. The entire unit is only 9 inches in diameter and 20 inches long and cuts weight from 800 pounds to 90 pounds.

Electric autos will probably have regenerative braking, with a flywheel taking up energy as the car is slowed and releasing it to accelerate again. This will reduce greatly the load on the battery and lengthen battery lifetime. Friction braking will be employed in a secondary mode. A special form of clutch is required in connection with the flywheel storage unit. The flywheel would be housed in an airtight chamber, from which the air is evacuated to reduce friction.

Basically, the electric car is simpler than its gasoline cousin, and it should be more dependable. The electric motor, which will be reversible, has ample torque at all speeds. No longer will the owner be concerned about such things as a starter, oil filter, carburetor, automatic choke, spark plugs, radiator, antifreeze, water pump, fuel pump, distributor, muffler and tail pipe, and the crazy-quilt mixture of anti-pollution devices now being superimposed upon this already complex bit of gadgetry.

The electric auto is ideally suited for urban-suburban use. Not only is it pollution free, but it wastes almost no energy in traffic. When the car is halted in traffic, the motor does not idle; it simply does not run. An electric auto accelerates easily from a stop to any speed within its limits and operates as efficiently at slow speeds (even at a crawl) as at highway velocities. Recently developed models accelerate as well as their gasoline cousins.

In the winter, heating for the passenger compartment is provided by a small alcohol-fueled unit. Cooling in summer is accomplished via a compressor that operates from the battery.

Special batteries such as sodium-sulfur, zinc bromide, and tubular plate lead-acid are being developed specifically for use in electric vehicles. They can be repeatedly discharged to 20 percent of full charge (80 percent depth of discharge) and recharged overnight. If the car is driven to work, the battery can be recharged while parked during the day. If the car is used for errands, the battery can be partially recharged during longer stops, if desired.

While lead-acid batteries impose a range limit of about 100 miles, within a few years the new batteries already under development will double and even triple the range.

The electric auto would be ideal today as a second car in the typical household: for urban driving in general and for commuting purposes. As the price of gasoline rises, the "E-car" would be used most of the time, and the gasoline model would soon become the backup vehicle. Cars under four years of age account for most of the total mileage today. Thus, new electric autos would be used more and, in a short while, would be saving gasoline very much out of proportion to their numbers.

Single-car families also could adjust to the electric auto without great inconvenience in most instances. Battery packs will be readily interchangeable to permit trips of several hundred miles in a day. Alternatively, a gasoline car could be rented for an occasional longer trip, and the owner would still be ahead economically.

The point is that the cost of operating an electric car will not go up very fast even if electricity costs rise somewhat. It could even decline as more efficient and less expensive batteries come on the market. On the other hand, gasoline models face the sure prospect of escalating fuel costs at the whim of foreign suppliers. In combination with photovoltaics, the electric auto should provide a transportation system that is about as inflation-proof as one could hope for. As the efficiency of photovoltaic systems and batteries increases, the unit cost of motive power will steadily decline—a sharp reversal from today's escalating fossil fuel prices.

Photovoltaics and the electric auto would provide a transportation system that is essentially pollution free. At the power plant end, expanded use of fossil fuels could be curbed, and additional nuclear plants would not be needed, while in the cities the largest source of air pollution—gasoline and diesel engines—would be replaced. The latter would provide a positive feedback mechanism: As smog decreased in our cities, more sunlight would come through, and photovoltaic systems would become more effective. Working together, the two technologies would be a double-edged sword against pollution.

Fortunately, the incremental cost of converting to electric autos would be minimal. Eighty percent of all cars are scrapped within ten years of their manufacture. The total truck and bus fleet has an average life of only seven years. Switching to electric cars is the most cost-effective way open to free this country from dependence on foreign oil. The net cost would be the research and development expense associated with "proving in" the new species.

The time has arrived to seriously consider a well-coordinated transition to electric vehicles over a reasonable period of time. This could be most efficiently accomplished through the normal process of development and commercialization, as for any other product. In the event of a severe emergency, however, a more rapid changeover could be effected, if the development work has been completed in advance. More effort by the private sector and the government should be put into the development of electric vehicles as well as photovoltaics.

The eventual benefits to the nation of an all-electric personal transportation system and an all-electric fleet of commercial light trucks, in combination with clean photovoltaic energy—the "PV-EV" option—would be enormous.

Supplies of liquid fuels need to be husbanded carefully to assure adequate amounts for those vehicles that truly require them, such as commercial aviation, defense equipment of many types, the machines for agriculture, and the like.

Under normal circumstances, electricity from the grid requires the burning of three BTUs of fossil fuel for each BTU of electricity produced. This is why experts such as Amory Lovins have not been enthusiastic about the electric auto. Under ordinary circumstances, electric autos would not be an appropriate use of energy and would reinforce the trend toward larger central station generating systems. However, if the additional power comes from photovoltaics, the picture changes completely, and the future looks bright indeed!

The two technologies—photovoltaics and electric vehicles—should be brought on in lockstep. At the present state of knowledge, both are entirely feasible, requiring only applied development.

In view of the fundamental fuel and pollution problems that plague our age, the presumption of helplessness is itself one of the most vexing parts of the whole energy situation. Pogo, in Walt Kelly's wise and gentle comic strip, had it figured out pretty well: "We have met the enemy," he said, "and he is us."

Photovoltaics seems to be free of the hazards, tradeoffs, and inequities fostered by fossil fuels. Photovoltaics can realistically accomplish the following:

- Ease the balance-of-payments account stemming from overdependence on foreign oil supplies
- Help reduce pollution caused by the burning of fossil fuels, especially when solar-charged electric vehicles become widely used
- Curtail long-term carbon dioxide pollution of the atmosphere
- Render unnecessary vast increases in coal mining
- Reduce the chance of renewed inflation stemming from possible future oil price escalation
- Buy time to resolve vexing problems associated with the siting and operation of nuclear-electric plants
- Provide a clear signal to the rest of the world of U.S. determination to become energy self-sufficient

Energy is so basic to this society—energy habits and policies are so deeply embedded in both physical surroundings and social values—that the structure of meaningful change in the energy market must arise from, or at least have the assured support of, a broad spectrum of the populace. In this society, energy supplies and basic issues of political and economic health, quality of life, environmental protection, and social equity should be complementary, not competitive, goals.

NOTES

1. Luther J. Carter, "Academy Energy Report Stresses Conservation," *Science* 207 (25 January 1980), p. 385; Eric Hirst and Bruce Hannon, "Effects of Energy Conservation in Residential and Commercial Buildings," *Science* 205 (17 August 1979), p. 656.
2. Conference of energy specialists at Stanford University, June 1980, as reported in *Science* 209 (11 July 1980), p. 246.
3. Report of the Energy Policy Task Force, Houston. November 5, 1980, Michael T. Halbouty, Chairman. See "Reagan Advisers Urge Phase-Out of Oil Profit Tax," *Washington Post,* 13 December 1980, p. A-1.
4. Richard A. Kerr, "Geopressured Energy Fighting Uphill Battle," *Science* 207 (28 March 1980), p. 1455: "It now appears that even modest hopes for geothermal energy will be difficult to fulfill."
5. S. C. Morris, P. D. Moskowitz, W. A. Sevian, S. Silverstein, and L. D. Hamilton, "Coal Conversion Technologies: Some Health and Environmental Effects," *Science* 206 (9 November 1979), p. 654.
6. Amory B. Lovins, "Energy Strategy: The Road Not Taken?" *Foreign Affairs* (October 1976), pp. 65–96. Lovins rests his case on the high cost of new coal and nuclear central station plants compared to the cost of energy produced and consumed locally.
7. Carrol I. Wilson, "Coal: Bridge to the Future." Report of the World Coal Study (Cambridge, Mass.: Massachusetts Institute of Technology, 1980).
8. Council on Environmental Quality, *The Good News About Energy* (Washington, D.C.: Government Printing Office, 1979), pp. 23–25.
9. U.S. Department of Energy, *Technology Characteristics—Environmental Information Handbook.* DOE/EV-0072 (Springfield, Va.: National Technical Information Service, 1980), p. A-3. Appendix A is an excellent summary of the environmental factors involved in operating a coal-fired power plant.
10. William S. Cleveland and T. E. Graedel, "Photochemical Air Pollution in the Northeast United States," *Science* 204 (22 June 1979), p. 1273.
11. [Jenny Hopkinson], "Tracking the Clues to Acid Rain," *EPRI Journal* 4 (September 1979), p. 20.
12. Bill Richards, "U.S. Conversion to Coal Would Add to 'Acid Rain' in Canada," *Washington Post,* 2 February 1980; Joanne Omang, "Acid Rain: Push Toward Coal Makes Global Pollution Worse," *Washington Post,* 12 December 1979; "Rain Falls Everywhere," *Washington Post,* 7 December 1980, editorial.
13. Philip H. Abelson, "Energy and Climate," *Science* 197 (2 September 1977), editorial.
14. Luther J. Carter, "A Warning on Synfuels, CO_2, and the Weather," *Science* 205 (27 July 1979), p. 376; Nicholas Wade, *Science* 206 (23 November 1979), p. 912.
15. W. S. Broecker, T. Takashi, H. J. Simpson, and T. H. Peng, "Fate of Fossil Fuel Carbon Dioxide and the Global Carbon Budget." *Science* 206 (26 October 1979), p. 409.
16. M. A. Khalil and R. A. Rasmussen, "Carbon Monoxide in the Earth's Atmosphere: Increasing Trend," *Science* (6 April 1984), pp. 54–56.
17. Philip Abelson, "Energy and Climate," *Science* 197 (2 September 1977), editorial. See also his editorial "Acid Rain," *Science* 221 (8 July 1983), in which he urges "the development of technologies that effectively reduce emissions while not creating additional environmental problems."
18. Eliot Marshall, "Synfuels Program Born in Confusion," *Science* 205 (28 September 1979), p. 1356; Ann Hughey, "Synfuels Versus the Political Realities," *Forbes* (20 August 1979), p. 41; Kathleen K. Wiegner, "The Water Crisis: It's Almost Here," *Forbes* (20 August 1979),

p. 56; [Nadine Lihach], "Water Water Everywhere But . . ." *EPRI Journal* 4 (October 1979), p. 7.

19. Richard L. Meehan, "Nuclear Safety: Is Scientific Literacy the Answer?" *Science* 204 (11 May 1979), p. 571; Eliot Marshall, "Assessing the Damage at TMI," *Science* 204 (11 May 1979), p. 594; Amory B. Lovins, "Invited Testimony for Hearings on the Costs of Nuclear Power Before the Environment, Energy, and Natural Resources Subcommittee of the Committee on Government Operations, U.S. House of Representatives" (21 September 1977), (photocopied).
20. "Political Problem," *Scientific American* (May 1980), pp. 78–82.
21. Robert Stobaugh and Daniel Yergin, eds., *Energy Future: The Report of the Energy Project at the Harvard Business School* (New York: Random House, 1979), p. 135.
22. Net oil imports and annual cost to the U.S. were (U.S. DOE *1982 Annual Energy Review*):

	Mill. Bbls. per Day	Cost
1980	6,365	$61.15 Billion
1981	5,401	60.88
1982	4,298	45.25
1983	4,249	45+

23. Lester R. Brown, *Running on Empty: The Future of the Automobile in an Oil Short World* (New York: W. W. Norton, 1979).
24. Stobaugh and Yergin, *Energy Future,* pp. 30–31.
25. Daniel Yergin and Martin Hillenbrand, *Global Insecurity: A Strategy for Energy & Economic Renewal* (New York: Houghton Mifflin, 1982).
26. International Energy Agency, *World Energy Outlook* (Paris: OECD, 1982). See also James Cook, "The Great Oil Swindle," *Forbes Magazine* (15 March 1982) pp. 99–108: "Whether it lasts six months or three years, the so-called oil glut is only a correction and not a change in the fundamental conditions of supply."
27. "Oil Embargo Might Cost $270 Billion, CBO Reports," *Science* 208 (6 June 1980), p. 1123.
28. Stobaugh and Yergin, *Energy Future,* p. 216.

FURTHER READING

Abelson, Philip H. "The Oil Price Spiral." *Science* 207, February 29, 1980, p. 937.

"Acid Rain Research" (Special Report). *EPRI Journal,* Palo Alto, Calif.: November 1983.

Bodansky, David. "Electricity Generation Choices for the Near Term." *Science* 207, February 15, 1980, pp. 721–728.

Clark, W. C., ed. *Carbon Dioxide Review 1982.* New York: Clarendon, Oxford University Press, 1982.

"Conservation and Solar Energy Programs of the Department of Energy—A Critique." Office of Technology Assessment, U.S. Congress. Washington, D.C., June 1980.

"Convention for a Solar America: Sense of Urgency on Energy Advocated for National Policy." *Solar Engineering,* October 1979, pp. 22–24.

"DOE Progress in Electric Vehicles." *EPRI Journal,* June 1984, pp. 32–34.

The Electric Vehicle Mini-Guide, 1980. Electric Vehicle Council, 1111 19th Street N.W., Washington, D.C. 20036.

Global Energy Futures and the Carbon Dioxide Problem. Council on Environmental Quality. Washington, D.C.: Government Printing Office, 1981.

Global 2000 Report to the President. Council on Environmental Quality and the U.S. Department of State. Washington, D.C.: Government Printing Office, 1980.

The Good News About Energy. U.S. Executive Office of the President, Council on Environmental Quality. Washington, D.C.: Government Printing Office, 1979.

"Hayden and Lovins Debate Institution Roles." *Solar Engineering,* October 1979, p. 24.

Longrigg, Paul. *Rapid Charging of Lead-Acid Batteries for Electric Vehicle Propulsion and Solar Electric Storage.* SERI/RR-742-1068. Golden, Colo.: Solar Energy Research Institute, 1981.

Lovins, Amory B.; Lovins, H. L.; Krause, F.; and Bach, W. *Least-Cost Energy—Solving the CO_2 Problem.* Andover, Mass.: Brick House, 1981.

Novick, Sheldon. *The Electric War—The Fight Over Nuclear Power.* San Francisco: Sierra Club Books, 1976.

Okrent, David. "Comment on Societal Risk." *Science* 208, April 25, 1980, pp. 372–375.

Stirewalt, E. N. "Get Set for Electric Vehicles." *Solar Age,* September 1981.

[Wayne, Mary]. "The Promise and Puzzle of Electric Vehicles." *EPRI Journal* 4, November 1979, pp. 7–15.

Whipple, Chris. "The Energy Impacts of Solar Heating." *Science* 208, April 18, 1980, pp. 262–266.

Zaleski, L., and Pierre, C. "Energy Choices for the Next 15 Years: A View from Europe." *Science* 203, March 2, 1979, pp. 849–851.

CHAPTER 9
THE SILENT REVOLUTION

Just as photovoltaic cells convert sunlight into electricity in total silence, so the emergence of photovoltaics as a commercially viable energy source is occurring quietly and with little fanfare. An energy revolution in the making goes virtually unnoticed by the general public.

Yet we are living in the midst of an incredible energy transition. Photovoltaics is serving the forces of decentralization, environmental concern, economic necessity, simplicity, and energy independence. Everywhere, people are using photovoltaics in dozens of ways, for the most part in small devices, but increasingly to meet larger power needs, too. Nearly 200 million persons own photovoltaic systems in the form of light-powered calculators or wristwatches. More than 20,000 persons worldwide live in photovoltaic-powered houses or vacation cottages. Thousands of farms are irrigated by photovoltaics. Millions of communication miles draw power from photovoltaics.

How fast is this revolution enveloping us, and where will it lead? In this chapter, we will attempt to paint an up-to-the-minute picture of the photovoltaics scene. The essence of the silent photovoltaic revolution can be stated in three powerful themes:

1. Photovoltaics works—silent and reliable, photovoltaics is cost-effective for many small applications as well as for large applications where utility electricity is not available or convenient.

2. The incredible technology base that exists in the universities, government labs, and industry is so strong that the cost reduction required to make photovoltaics a pervasive source of electricity will occur in the next ten years. We will set forth the details of this cost reduction, identify the key elements that must be realized in order for it to happen, and forecast the reduction as a function of time.

3. Given this cost reduction, the U.S. photovoltaic market can develop into a multibillion dollar industry by 1995. We will discuss why it will develop and who will benefit.

It should be clear that the photovoltaic revolution cannot be stopped—photovoltaics *will* become a pervasive source of clean, reliable electricity.

Photovoltaic systems are now producing more than 50 megawatts of power worldwide. In order to illustrate the momentum of the photovoltaic revolution, we will describe several of the key U.S. applications. It is convenient to break the market down by system size as we describe applications ranging from those that produce fractions of a Watt of electrical power (milliwatts) to those that produce millions of Watts (megawatts).

CONSUMER PRODUCTS: MILLIWATTS TO ONE WATT

The revolution is being spread by the over 200 million persons who have photovoltaic systems in their purses, pockets, or desks! These systems operate by using room light or sunlight for power. Photovoltaic calculators are the most widespread example. They have captured over 60 percent of the free world's calculator market. Between 60 and 70 million are sold each year.

Light-powered calculators embody many attributes that photovoltaic systems must have in order to precipitate an energy revolution. The calculators are:

- Powerful and reliable, lasting for tens of years instead of the months or weeks that typical throw-away batteries last. Moreover, the photovoltaic calculators are as small as conventional calculators because the photovoltaic cells they use are highly compact.
- Convenient (because there are no batteries) and affordable. In fact, photovoltaic calculators fulfill these requirements so well that they have led many people to purchase calculators for the first time, thus increasing the world market.
- Very profitable. When Japanese companies announced photovoltaic power packs for calculators in 1978, they charged one dollar per calculator. Each power pack puts out about 1/50 of a Watt, so the price was a profitable $50 per Watt. Sanyo, Fuji, and Sharp entered photovoltaic production with a profit and were able to fund further research from their profits and government contracts. It is very important that the photovoltaic industry be profitable in order to convince management to support research and to acquire new production equipment, which can reduce prices and lead to much larger markets.

- Products that capture people's imaginations, leading to new ideas and new products. The success of the calculators has led to a proliferation of other photovoltaic-powered consumer applications. Photovoltaic-powered portable radios are just beginning to see significant sales. Probably one of the most successful consumer electronic products to be introduced by the Japanese electronic industry was the quality portable stereo, including tape deck—it was introduced as the Sony Walkman. In 1980, after major success with photovoltaic calculators, Japanese companies introduced portable photovoltaic power packs to go with the portable radios. Several hundred thousand have been sold, and with lighter packaging and lower power requirements of the radios, we expect to see millions of portable radios powered or recharged by photovoltaic one-half-Watt chargers. The range of consumer products that can be operated on less than one Watt is extremely large. A few that are in prototype or are available from various catalogs on the market include: photovoltaic-powered portable televisions, toys, music boxes, telephones, radio-controlled model aircraft, thermometers, and humidistats. The products range from the practical to the entertaining. A photovoltaic-powered pith helmet uses a patented photovoltaic fan and wet sponge, thus providing evaporative cooling. Nearly 50,000 of these novelty helmets have been sold.

ONE WATT TO TEN WATTS: FENCES TO MOTOR BOATS

This sector of the photovoltaics market is where we graduate to applications involving a small amount of power to do significant work in remote places where we would normally use batteries. The applications are incredibly diverse, and once photovoltaics is used, the customer chooses it again because of its silence, portability, low cost, reliability, and low-maintenance operation. We are aware of several hundred different applications of this type that now use photovoltaics instead of other energy sources. They include the following:

- Nearly 100,000 electric fence chargers. The Solarex Parmak photovoltaic-powered fence charger is even in the Sears farm catalog. Such fence chargers safely restrain cattle, hogs, and other animals.
- Nearly 10,000 solar water heaters having pumps driven by 10-Watt photovoltaic panels. The panels are more reliable than utility power with switches and timers. Virtually all solar water heater manufacturers in the United States have developed systems that pump the water through the panels to the storage tank using low-Wattage photovoltaic panels and small direct-current, motor-driven pumps. For example, the Pulstar

Figure 9-1 Daytime sun becomes nighttime light. Photovoltaic panels generate electricity that illuminates this highway sign. The electricity is generated during the day and stored in batteries; the highway sign is illuminated after sunset. (Courtesy: Kyocera)

system pumps water with power from a small photovoltaic panel (less than 10 Watts), thus saving panel costs. Photovoltaic-powered pumps work without any controls or switches. When the sun is hot enough to heat the water, solar energy causes the panels to put out enough electricity to slowly pump the water through the heat collectors. As the sun gets hotter, the pumps work harder, and the hot water systems pump their solar-heated water into storage. This application is a perfect example of "making hay while the sun shines." The pumping speed precisely matches the "load" as the amount of sunlight rises and then falls again. We expect that soon nearly all solar hot water systems will have photovoltaic pumps, and that this successful "sun-follower" application will lead to many other markets.

- Thousands of remote signals, lights, etc., powered by small photovoltaic panels. One photovoltaic application with worldwide sales, which has surprised many analysts, is the use of photovoltaics to charge a battery during the day, then use the stored electricity to power a light at night (Fig. 9-1). Thousands of photovoltaic-powered lights have been installed around parking lots, along walkways through parks, as emergency lamps, etc. Typically, a 12-Volt, deep-discharge battery is charged by a 20- to

Figure 9-2 PV to go. Photovoltaic arrays needn't be huge. This compact array folds up for storage in a 6 inch by 12 inch pouch. Small, portable arrays of this kind provide electricity for use on camping trips and other remote-site activities. (Courtesy: Solec)

40-Watt photovoltaic panel. On a bright day, 200 Watt hours are stored, permitting a 40-Watt fluorescent light to operate for 5 hours. Obviously, larger panels would permit more powerful lights to be used for longer periods of time. At many sites, photovoltaic-powered lamps are the most economic, reliable way to light the night because the cost of running an electric line for a hundred feet or so can cost more than $1,000.

- Thousands of motor boat batteries and sailboat batteries, trickle-charged by photovoltaic panels. The panels keep the batteries topped up so that bilge pumps and other devices don't run them down. Starting motors always work with photovoltaic-powered boat batteries.
- Battery packs for many consumer products (including TVs, radios, stereos, and microcomputers), recharged by 10-Watt photovoltaic panels. The packs can also provide a hiker or camper with portable power for radios and emergency beacons. Many companies offer lightweight, rugged power packs so that reliable electricity can be available wherever people go. For example, Solec, a photovoltaic module manufacturer in Hawthorne, California, produces a complete line of portable power modules that provide on-the-spot electricity for hunting, fishing, camping, backpacking, and mountain-climbing expeditions. The "solar chargers" are available in 2.5-pound, 9-Watt and 9.4-pound, 30-Watt versions. The Solec "Man Pack Array" is a five-section panel that folds up and can be stored in a 6 inch by 12 inch pouch. (Fig. 9-2).

Figure 9-3 The Rob Robinson house (shown on our cover) near Santa Barbara, California, employs a photovoltaic array for basic energy needs. Because the house is not connected to a utility, batteries are used to provide 10 kilowatt hours of stored electric power. Such "off-grid" uses are increasingly popular. (Courtesy: Solarwest Electric)

REMOTE POWER: 100 WATTS TO 1,000 WATTS (1 KILOWATT)

Photovoltaic systems in this size range, costing from $1,000 to $20,000 installed, are beginning to serve the power needs of hundreds of consumers and industries where reliability is a key factor. These applications would normally get their electricity from small gasoline or diesel engines or large deep-cycle rechargeable batteries.

One-hundred-Watt photovoltaic systems can power loads such as these:

- Communication systems that draw 50 Watts for 2–4 hours per day
- Forty-Watt fluorescent lights that operate 10–12 hours each night
- One-hundred-Watt ham radios that operate 3–5 hours a day
- Five-hundred-Watt emergency signal systems that operate for 20–30 minutes per day

Five-hundred-Watt systems can handle larger applications. One of the most exciting 500-Watt systems is the Rob Robinson house shown on our cover (Fig. 9-3). Rob and his wife purchased a lot with a view, in the mountains near Santa Barbara, California. When they decided to build their home in 1979, they went to Southern California Edison (SCE) to ask for an electrical hookup. The power lines were less than 1,000 feet from their site. However, the community had zoned their lot for "underground" electricity distribution. This meant that the power lines had to be brought 1,000 feet through solid granite. The California Utilities Commission had granted SCE the right to recover the costs of the underground line. SCE quoted the Robinsons a $5,000 charge to hook up their house, so the couple decided to use photovoltaics instead. Rob designed the 420-Watt system and built it for $10,000. He took advantage of federal tax credits ($4,000) and California tax credits ($2,000), which meant his out-of-pocket cost was less than the SCE hookup charge! The modest electrical loads in this modern house are all photovoltaic-powered. The photovoltaic-generated electricity is stored in a deep-discharge battery bank, which holds three to four days' of storage. The system uses direct current for lights and alternating current for small motor-driven appliances. Rob and his wife cook, refrigerate, and heat with bottled propane. Thanks to photovoltaics, they have not had an electric bill in five years. Rob has now formed Solarwest Electric in Santa Barbara, California. As an ARCO Solar distributor, he has sold and installed several hundred photovoltaic-powered houses and a thousand small, 100-Watt to 1-kilowatt systems through his network of sales representatives. Rob estimates that he is now selling 200–400 new systems each year.

One-kilowatt photovoltaic systems are at the top of the scale for small systems. They can power:

- One-horsepower pumps that operate 5–6 hours per day
- Small homes with lights, television, radio, a well pump, and some appliances

Figure 9-4 Monhegan Island, in Maine, has a surprisingly large number of photovoltaic systems. Many homes on the island use PV to supplement diesel- or gasoline-powered electrical generators. Even the local post office is PV-equipped. (Courtesy: David Sleeper)

- Medical refrigerators that make ice and cold water
- Signals that put out 200 Watts for 24 hours a day

There are about 20 photovoltaics distributors serving the remote-power market. We estimate that 5,000–10,000 small, less-than-one-kilowatt systems are now sold in the United States annually for applications where a small amount of electricity is needed to power an important load. Before photovoltaics was available, these needs were served by batteries, gasoline-powered generators, or diesel-electric generators. Batteries are very expensive, with electricity costs of $1–$2 per kilowatt hour. Generators are noisy, they pollute, they require maintenance and fuel, and the electricity they produce costs from $0.50–$2.00 per kilowatt hour. Depending on climate and storage, photovoltaic electricity ends up costing about the same, but the aggravations of noise, pollution, frequent maintenance, etc., are done away with.

Hundreds of small (35- to 1,000-Watt) off-the-grid photovoltaic systems have been installed in an unlikely spot—Maine. The photovoltaics market there was stimulated by a decision made by Central Maine Power Company to charge $6,000 to install a power line for new service for any distance up to 2,000 feet. Very much like Rob Robinson, many families opted for photovoltaics and no monthly electricity bills rather than pay the $6,000 for a utility hookup. Many remote sites and offshore islands previously had oil or propane-fired generators. Photovoltaic systems offered silent, non-polluting electricity despite the comparative lack of sunshine. The small community on Monhegan Island (12 miles off the coast) has become famous for photovoltaic houses. Over 50 percent of the homes there have some photovoltaics to replace or supplement generators or kerosene lamps. Monhegan even has the nation's first solar-powered U.S. Post Office (Fig. 9-4). Sandra Dickenson, one of the first Monhegan Island photovoltaic homeowners, is now a photovoltaics consultant and has moved to Port Clyde on the mainland. Maine's Office of Energy Resources has assisted the movement to photovoltaics with workshops and a pamphlet, "Electricity from the Sun." Five Energy Extension Service agents conduct workshops and offer slide shows throughout the state.

It would be easy to fill this book with descriptions of the thousands of small photovoltaic systems in the Sunbelt. We think the Maine projects show that small, remote photovoltaic systems can work effectively wherever a house has no existing hookup with the utility grid, and a relatively modest amount of electricity is sufficient.

ONE KILOWATT TO TENS OF KILOWATTS: HOUSES, FARMS, VILLAGES

This market is just beginning to develop. A number of people have bought photovoltaic systems of this size even though utility electricity would have been less expensive. The customers include well-to-do individuals who want a house that is energy conserving, has active and passive solar systems, obtains its electricity from renewable sources, and has a sufficient electrical supply to power all the appliances found in a typical American upper-middle-class home. Commercial customers are also found in this size range. Farmers who need reliable, low-maintenance power for homes, irrigation, and stock watering are purchasing photovoltaics. Investors are purchasing 2- to 10-kilowatt systems from companies like United Energy Corp. because they believe in renewable energy and photovoltaics. They prefer to risk their

Figure 9-5 This photovoltaic home was designed for the seemingly inhospitable region of the Hudson River Valley in New York, where clouds often obscure the sun and up to 100 inches of snow fall each year. The home is not connected to any utility company. (Courtesy: Solar Design Associates)

money on a high-technology electricity source rather than on an oil well or feeder cattle, race horses, etc.

We will discuss only a few of the many hundreds of systems operating in this size range. Our apologies go to those superb projects that are not included.

Frank Arenas, a Tampa, Florida, builder/developer, is using the Solara II house as a luxury prototype ($280,000) to design and construct smaller all-solar homes and town homes in the $60,000–$80,000 range. Solara II has 1.2 kilowatts of Solarex semicrystal photovoltaic modules, solar water heaters (which heat water for the swimming pool as well as for the home's water taps), and a unique household cooling system that circulates air through a double floor slab. The house is not connected to the utility.

Laurence Jennings, President of Photron, Inc., in Willits, California (a manufacturer and distributor of alternative electrical power systems and balance-of-systems components, specializing in photovoltaic/engine generator hybrid power), designed a unique home for a Northern California contracting firm. Installed in what has come to be called "The Pyramid House," the system includes a 1.5-kilowatt ARCO Solar photovoltaic array mounted with Photron's unique "theft resistant" array structures and a 6.5-kilowatt backup Onan generator coupled to a 5-kilowatt Photron Gencharger (a specially built, high-capacity generator-driven battery charger). There is also a sophisticated Photron "BRAIN" (battery state-of-charge controller), which brings the backup charging system on and off line automatically at any battery state-of-charge the customer desires, 5 kilowatts of (dual) DC to AC inverter capacity, 36 kilowatt hours of battery storage capacity, and full remote digital monitoring of the entire system. The system provides power for a wide variety of exotic loads including waterfall pumps (there is a waterfall inside the house), Jacuzzi, microwave oven, dishwasher, trash compactor, and a shop full of tools. The local utility would have charged a phenomenal $25,000 to extend a power line all the way to this 9,000-square foot, solar-heated and -cooled home. Because the homeowner avoided these high connection charges, the photovoltaic system was immediately cost effective.

The use of photovoltaics to power totally a large, comfortable house in the Hudson River Valley in New York (100 inches of snow each winter and relatively little sun) was indeed a challenge. Steven Strong of Solar Design Associates in Lincoln, Massachusetts, was chosen to design and build the house (Fig. 9-5). It is a full-size photovoltaic-powered residence without any connection to a utility. The house incorporates direct passive solar gain, superinsulation (R30 and more), special Heat Mirror windows, and active solar water and space heating. A bank of storage batteries provides several days of electricity storage.

Steven Strong has often referred to the market for photovoltaic homes as a "trickle down/trickle up" market. Young families of modest means are purchasing very small, two- or three-panel photovoltaic systems to provide one to two lights and music, while very wealthy families—even those removed from the Sunbelt—have purchased very expensive, grid-independent, energy-conserving photovoltaic houses. From 5,000 to 10,000 families now get all of their electricity from the sun. Some systems are fully economic, and all are statements of the owners' energy consciousness. As we write this chapter, several thousand photovoltaic systems for large and small homes are being designed and built.

THE GRID-CONNECTED PHOTOVOLTAIC HOUSE

There are about 40 houses in the United States that have photovoltaic systems while also being connected to the utility grid. In these situations, the utility acts as a source of backup power and pays the photovoltaic system owner for excess electricity generated by the system when the sun is bright and the house load is low: The system then sends electricity into the grid, in effect turning the tables on the utility by selling power to it.

We have already discussed the pioneering grid-connected houses, including the John Long House in Phoenix, the Carlisle House in Boston, and the three Residential Experiment Stations funded by the Department of Energy.

The Tennessee Valley Authority (TVA), which generates most of its electricity by hydropower from dams on the Tennessee River, is experimenting with residential photovoltaics as another renewable source of electricity. Several photovoltaic houses have been built in the TVA region. For example, Solenergy Corp. of Woburn, Massachusetts, and Development Sciences of Sagamore, Massachusetts, designed and built the roof-integrated, 4-kilowatt system at the TVA Power Service Center in Chattanooga, Tennessee. (Solenergy is the only privately held photovoltaic module manufacturer in the United States. It was formed in 1977 by Bob Willis, a pioneer in the photovoltaics industry.

Besides TVA, several other utilities have cooperated in building grid-connected photovoltaic houses. These include San Diego Gas and Electric, Arizona Power and Light, Georgia Power Company, Florida Power and Light, and Southern California Edison.

The Boston Edison Impact 2000 house in Brookline, Massachusetts, has enabled nearly 100 million persons to learn about photovoltaics, conservation, and load management. During late 1983 and early 1984, the construction and design of the house was chronicled by the Emmy Award-winning television series, "The All New This Old House," which aired nationally on over 200 public television stations. Steven Strong of Solar Design Associates in Lincoln, Massachusetts, designed the house for Boston Edison. Impact 2000 serves as a laboratory for the company's first photovoltaic unit, which generates 4 peak kilowatts. The house also features direct passive solar gain for winter heating, superinsulation, a solar hot water system, a ground-coupled heat pump, an integrated solar sunspace, and a 4-kilowatt roof-integrated array of photovoltaic modules made from silicon ribbon from Mobil Solar Energy Corp. The photovoltaic system is backed up by the Boston Edison grid.

Impact 2000 features new technology coupled with traditional touches such as fireplaces. Thus, the house provides the best of both worlds: energy efficiency and comfort. Extensive monitoring devices will provide component suppliers, engineers, builders, bankers, and homeowners with useful information on the house's performance. Impact 2000 has literally dozens of clever innovations with its extensive use of conservation, passive solar, active solar hot water, and the like. We will now highlight some of the key attributes of the house's photovoltaic system.

Twenty-four large 180-Watt, 6-foot by 4-foot Ra 180 Mobil Solar modules are installed on the rafters. They not only provide 60–90 percent of the house's electricity, they also serve as the roof of the house.

The photovoltaic panels feed a 4-kilowatt inverter made by American

Power Conversion Corporation (from the Boston area). The inverter changes the photovoltaic direct current to 60-cycle, 110-Volt house current. A control unit then directs this electricity to the house lights and appliances. If the solar electricity generated exceeds the house load, the control unit "sells" the excess to Boston Edison for use by other families on the block. When the power supplied by the sun does not equal the electrical needs of the house, the control unit automatically directs the Boston Edison-generated electricity to the house.

The system's inverter provides lightning protection for the house's electrical system. And if the utility company has a power failure, the control unit automatically isolates the house's photovoltaic system so that power from the solar panels does not bother a line service person who is not expecting to encounter current flowing out of a customer's house. As with all photovoltaic systems, the Boston Edison Impact 2000 home meets all aspects of the 1984 National Electrical Code.

In the history of photovoltaics, the "All New This Old House" may well educate more persons as to the value of photovoltaics, conservation, and active and passive solar energy than any other single event.

LARGE INDUSTRIAL/COMMERCIAL SYSTEMS—10 KILOWATTS TO 1 MEGAWATT

This sector of the market is still new. As we indicated previously, the U.S. Department of Energy supported several projects in this size range and proved that photovoltaic systems could be used to power schools, hospitals, shopping centers, museums, and other large establishments. However, the private sector has built only a few systems of this size. Photovoltaic electricity costs are still sufficiently high that the U.S. market has been slow to develop.

The most interesting project in this size range is the 200-kilowatt manufacturing plant built by Solarex in Frederick, Maryland, discussed earlier.

10-MEGAWATT ENERGY FARMS OF 2.5-KILOWATT CONCENTRATORS

United Energy Corporation of Foster City, California, has built and sold large multimegawatt systems using the federal renewable energy tax credits, accelerated depreciation, and the California solar energy tax credit.

United Energy is owned and operated by Ernie and Del Lampert, who have built this company into the second largest photovoltaics manufacturer in the world. They did this by combining photovoltaics and solar hot water in

Figure 9-6 The Intercultural Center at Georgetown University in Washington, D.C., receives electricity from a 300-kilowatt photovoltaic array mounted on its roof. (Courtesy: Solarex Corp.)

Figure 9-7 This system generates both electricity and hot water. Lenses concentrate the sunlight to 100 times its original intensity. Solar cells then produce electricity, and water circulating through a pipe system is heated to 140°F. (Courtesy: United Energy Corp.)

Figure 9-8 The "Energy Farm" in the Borrego Springs desert of California provides electricity and heat for an advanced ethanol production plant. (Courtesy: United Energy Corp.)

one product. They have sold over 5,000 2.5-kilowatt hybrid units to small individual investors wishing to take advantage of solar tax credits.

A photovoltaic electricity-hot water hybrid system built by United Energy uses a point-focus Fresnel lens concentrator (Fig. 9-7). The sunlight is concentrated by a factor of 100 onto a single-crystal solar cell, which converts the light to electricity. The cell is mounted on a copper pipe; water flowing inside the pipe extracts heat at 140°F. Each module is designed to produce 2.5 peak kilowatts of electricity (20 kilowatt hours per day) and 40,000 Btu's of hot water per hour. United Energy uses two-axis tracking: Each

Figure 9-9 The Barstow, California, "Energy Farm" consists of 2.5-kilowatt photovoltaic modules positioned in 6-inch-deep ponds. The modules—100 in each pond—track the sun. (Courtesy: United Energy Corp.)

module pivots on a horizontal axis while the platform carrying 100 modules and floating on a pond of water rotates slowly around a central pivot. United Energy uses 3M Fresnel lenses and Applied Solar Energy cells, as well as cells it produces in its own plant.

At United Energy's "Energy Farm" in the Borrego Springs desert near San Diego, each circular pond is 246 feet in diameter, contains 100 2.5-kilowatt modules, and provides 0.25 megawatts of electricity under peak load conditions (Fig. 9-8). The total system produces 3-5 megawatts, which power a facility that produces ethanol for use in motor fuels.

A second plant at Barstow, California, is very similar to the one at Borrego Springs (Fig. 9-9). Again the 250-kilowatt ponds are connected to produce megawatts of power. Each investor in the plant owns a specific module and receives a check each month for the energy produced. Both systems—Barstow and Borrego Springs—have contracts with the local utilities under which they sell excess electricity to the utilities when the ethanol facilities are shut down or do not need the full capacity of the photovoltaic plants.

LARGE CENTRAL STATIONS: 1 MEGAWATT TO 100 MEGAWATTS

Under normal market conditions, megawatt-size photovoltaic systems would not be ordered until their cost approached that of conventional sources of electricity. Most analysts find that photovoltaic systems are not "economic" until installed costs drop to $3 per peak Watt for a complete system. However, the existence of tax credits has resulted in nearly 20 megawatts of central station photovoltaic systems being built. Congress passed a solar tax credit of 15 percent in addition to the normal 10 percent investment tax credit; owners of solar systems could use the 5-year accelerated

Figure 9-10 The world's largest photovoltaics system, as of 1985. This 6-megawatt system in San Luis Obispo, California, produces enough electricity for 200–400 homes. (Courtesy: ARCO Solar)

depreciation schedule instead of the 20-year or longer depreciation for a conventional generator. The state of California also passed a 25 percent state tax credit. The combination of the federal and California credits made it economic for ARCO Solar to "sell" a large photovoltaics plant to its parent, Atlantic Richfield. This resulted in the world's first major photovoltaic central station project, a 1-megawatt station at Hesperia, California, which uses flat-plate modules on large two-axis tracking pedestals. The power from the station is sold under contract to the local utility company. (All major California utilities—San Diego Gas and Electric, Pacific Gas and Electric, and Southern California Edison—offer long-term power purchase agreements for renewable-energy electrical generation plants.)

After completion of the Hesperia station in 1982, ARCO Solar started an even larger project, a 6-megawatt power plant at San Luis Obispo, California. Its two-axis tracking system uses reflectors to enhance the sunlight falling on its flat-plate modules (Fig. 9-10). Both ARCO Solar plants were constructed in less than a year, operate essentially unattended, and have produced reliable electricity with little or no maintenance. As their success demonstrates, there is absolutely no doubt that central station photovoltaics works.

Massive deployment will come when photovoltaic system prices drop from today's $10 per peak Watt installed to $3. We forecast this to occur within the next ten years.

Another major central station project was approached from a different point of view. The Sacramento Municipal Utility District (SMUD) decided they needed 100 megawatts of new capacity by 1993. After calculating the cost of all options—including oil- and gas-fired turbines, new coal capacity, and hydro—they concluded that photovoltaic power at $3 per installed Watt would compete with all other options. SMUD approached the U.S. Department of Energy and the California Energy Commission and said in effect, "We will buy a 100-megawatt photovoltaic system for $300 million. If you will pay the difference between present costs and $3 per Watt, we will build 1 megawatt in 1984 and increasing megawattage in subsequent years until 100 megawatts are on line." The Department of Energy demurred. However, Congress added $7 million to the Department's budget, and earmarked it for SMUD. The State of California provided about $2 million. The first 1-megawatt phase was built for $12 million. Phase Two, which was funded by Congress (again without DOE approval), will be operating in 1985 and will cost about $9 per Watt. SMUD is committed to going all the way to 100 megawatts, provided the gap between $3 per Watt and actual cost continues to be funded by the state and/or federal government.

Another project stimulated by tax credits is being planned by Solarex. Like ARCO Solar, Solarex plans to "sell" a megawatt plant to its parent, Standard Oil of Indiana. It will be built in California.

MONEY AND TAXES

Tax credits have played a large role in the development of photovoltaics. The reader should remember that total world sales of photovoltaics amounted to only 5 megawatts in 1981. Today, over 20 megawatts of central station photovoltaics have been engendered by tax credits and by the requirement that utilities purchase electricity from a photovoltaic plant at "avoided cost" rates. The impact of tax credits on photovoltaic prices and photovoltaic markets is highlighted below.

The photovoltaic revolution is happening quickly, effectively, and silently. The momentum created by the milliwatt-to-megawatt scenario just presented is expected to continue until photovoltaics finds its proper place in the nation's energy arsenal. The central question about the revolution is how fast will the use of photovoltaic cells continue to expand. In turn, this leads to a second question: How fast will the price come down?

These two questions are tied together, for declining prices boost sales, while market growth tends to bring prices down. The probable future costs of modules made by the five leading processes can be projected with reasonable assurance based on knowledge of the processes and recent price history (flat-plate modules using silicon are obtained by four procedures; concentrators that typically use small single-crystal silicon or gallium arsenide cells are obtained by the fifth procedure).

Likely module costs[1] are given in Table 9-1.

Table 9-1 Costs of Photovoltaic Modules, Dollars per Peak Watt (expected module efficiencies are given in parentheses)

Type of Module	*1985*	*1990*	*1995*
Single-crystal silicon	\$6.50 (11%)	\$4–\$5 (15%)	\$3 (16%)
Silicon ribbon or sheet	\$7.50 (10%)	\$3.30 (11%)	\$2–\$3 (14%)
Cast silicon ingot (polycrystalline)	\$7 (11%)	\$3–\$5 (13%)	\$3 (15%)
Amorphous silicon	\$5.00–\$6.50 (5%)	\$2–\$3 (8%)	\$1.70–\$2.50 (10%)
Concentrator modules	\$5–\$6 (14%)	\$3.30–\$4.00 (17%)	\$2.50 (20%)

The fact that several routes are open to the low-cost product of tomorrow lends credence to the general price decline reflected in these projections.

As a matter of interest, the cost of the materials used in making flat-plate modules presently is as follows:

Silicon (semiconductor grade selling at \$60 per kilogram)	\$1.60/peak Watt
Glass	\$0.07/peak Watt
Frame	\$0.11/peak Watt
Ethyl vinyl acetate sealant	\$0.11/peak Watt
Other components (dopants, metallization, etc.)	\$0.11/peak Watt
Total	\$2.00/peak Watt

The remainder is labor and other manufacturing costs and profit.

The potential for cost reduction inherent in the amorphous silicon process, which uses only 1/50–1/100 the silicon consumed in the single-crystal or polycrystalline process, is apparent. Its principal shortcoming is lack of stability; by 1995, a warranted useful life of only five years is projected.

Nevertheless, the amorphous silicon approach is considered the most promising in the long term for mass-market use. All major companies are pursuing it—on a contingency basis, if not as their principal method. U.S. industry teams involved include Energy Conversion Devices, Sohio, Chronar, ARCO Solar, Solarex, Spire Corp., AFG Glass Corp., and Polaroid.

By contrast, there is less potential for major cost reductions in single-crystal cells. Costs are reduced by changing from round cells to square ones and increasing cell efficiency, which will increase module efficiency from 11 to 15 percent. These changes can bring the cost of flat-plate, single-crystal modules down to $3 per peak Watt. The greatest advantage from the user's standpoint is assured long array lifetime. U.S. manufacturers of flat-plate modules include ARCO Solar, Solarex, Solec International, Motorola, and Solenergy.

Ribbon/sheet processes are cost competitive primarily because they conserve silicon—enough to overcome somewhat lower efficiency and limited useful life. Increased ribbon pulling rates and better process control can keep this approach viable for another decade. Four companies are developing it: Mobil Solar Energy, Westinghouse, Energy Materials Corp., and Arthur D. Little.

About 30 percent of today's modules are made from cast rectangular semicrystalline ingots of pure (semiconductor-grade) silicon which are readily sliced into rectangular cells. The principal advantage is that casting these cells is less expensive than growing single-crystal ingots, albeit semicrystalline cells are slightly less efficient. The casting process is also believed to tolerate lower-quality silicon, thus reducing cost further. Larger, more efficient plants will permit steady price reduction comparable to that for other processes. Companies involved include Solarex and Crystal Systems, Inc.

Concentrators will remain competitive because the cells they use are small, and they can take advantage of advancements in cell technology even though the cost of new cells may be high on a square-foot basis. Commercial module efficiencies of up to 20 percent are projected for 1995. Though restricted to regions with high direct insolation, concentrators achieve elevated operating temperatures, giving them a distinct advantage where the thermal energy can be put to good use. U.S. manufacturers include Intersol, Entech, and United Energy Corp.

As indicated earlier, when module costs reach the level of $3 per peak Watt (installed system cost of $5 a Watt), the range of uses will be greatly expanded, because photovoltaic power will become competitive with electricity from the grid in many areas. If interest rates are low enough, even the U.S. grid-connected residential market will begin to open up at this level.

It now appears that the world market for photovoltaics will expand during the next ten years as shown in Table 9-2, assuming that the nominal average installed system cost will fall from $9 per peak Watt in 1985 to $5 in 1990 and then to $3 in 1995 (or $2 if U.S. tax credits are extended).

Table 9-2 World Photovoltaics Market (annual sales, peak megawatts)

	1985	*1990*	*1995*
Without Extension of U.S. Tax Credits			
World Market	50	200	400
U.S. Market	20	100	200
With Extension of U.S. Tax Credits			
World Market	50	500	2,000
U.S. Market	20	300	600

If U.S. tax credits are extended, the cost of photovoltaic systems will drop, which in turn will stimulate the world market.

Within the United States, the markets for photovoltaic systems may be distributed as shown in Table 9-3.

Table 9-3 U.S. Markets for Photovoltaic Systems (total annual sales)

Market	*1985*	*1990*
	Peak megawatts of generating capacity	
Consumer products and battery chargers (up to 500 Watts)	5	30
Off-the-grid residential (200 Watts-2 kilowatts)	3	20
Off-the-grid commercial (1 kilowatt-500 kilowatts)	0.8	4
Grid-connected residential (2-5 kilowatts)		
-Without tax credits	0.2	1
-With tax credits	(0.2	3)
Grid-connected commercial including third-party-financed central stations (10 kilowatts-100 megawatts)		
-Without tax credits	10	50
-With tax credits	(10	240)
Government purchases (DOE, DOD)	1	5
Total	20	110
(With Tax Credits	20	302)

Nothing inhibits the further growth of the photovoltaic consumer products market except its eventual saturation at 75–100 megawatts per year. It will continue to provide a profitable, stable base but will not constitute a major market force due to the small size of the products.

Off-the-grid residential uses include vacation cottages, motor homes, house boats, mobile trailers, recreational vehicles, and small permanent residences and villages. Typically, one or two photovoltaic panels provide DC power for fluorescent lights, a radio, a music system, a water purifier, a small television set, and possibly a small well pump. Limited battery storage is included. Frequently, an adjustable array is used, otherwise the panels are mounted facing south at ±10° latitude. An inverter, which changes the direct current to alternating current, is often added for larger units (300 Watt–1 kilowatt) to power deeper well pumps, a vacuum cleaner, small tools, etc. If a diesel or gasoline generator was already in place, it might be retained to form a hybrid system in which it could be used for heavier loads such as washing clothes. The potential market is significant. An estimated 1 million houses in the United States do not have electricity. Three million motor homes and mobile trailers are candidates, as are some 2 million lived-in boats. Tents used by millions of campers are a further market.

Principal off-the-grid commercial uses include irrigation, water purification, remote signals of various kinds, communication relays, and cathodic protection of steel structures such as bridges. This sector will experience continued steady growth.

Grid-connected residential uses are now limited to a few dozen houses. Several were funded by the government for investigative purposes, while a few have been built by interested individuals. A number of prototypes are incorporated at Residential Experiment Stations in Massachusetts, Florida, and New Mexico. Excess power is sold back to the grid at the "avoided cost" to the utiliity, as required by the Public Utility Regulatory Policies Act (PURPA). As pointed out in Chapter 5, the greatest long-term impediment to the growth of this market sector is continued high interest rates.

The grid-connected commercial sector, primarily third-party-owned central station units selling power to utilities in California, is extremely sensitive to the availability of tax credits.

A major component of the projected world market depends upon replacement of diesel generators, primarily in Third World countries, as shown in Table 9-4 (the assumption is that the installed cost per peak megawatt will fall from \$9 in 1985 to \$5 in 1990 and to \$3 in 1995). The diesel-replacement market in the United States represents about 20 percent of the world market. The overseas market is extremely sensitive to photovoltaic prices, the price of diesel fuel, and the available credit in developing countries.

Table 9-4 Replacement of Diesels—Annual Sales (peak megawatts of generating capacity)

	1985	*1990*	*1995*
5 kilowatts and under	10	25	50
20 kilowatts and under	—	75	375
Up to 100 kilowatts	—	—	800
Total	10	100	1,225

CONCLUSION

Photovoltaics is friendly. It has myriad niches to welcome it. Once it finds its way into these niches, there will be no turning back: The market will gradually expand in dozens of directions, and each sector will grow in depth. Slowly but relentlessly, photovoltaics will replace today's conventional primary fuel sources.

Within the next ten years, photovoltaics will offer an economically feasible alternative to coal and nuclear power, yet today this simple fact goes virtually unnoticed by the general public. Even informed opinion is apparently unaware of the potential of photovoltaics, or is skeptical about its future. For example, an editorial in a leading paper,[2] citing the basic difficulties of both nuclear and coal generating plants, concludes:

> If Americans think they will need nuclear power in the 1990s, the government is going to have to intervene—and it may well have to build the plants itself. Alternatively, if Americans want to dispense with nuclear power they will need to find ways to burn more coal without driving up the death rates.

This is the core of America's present electric power dilemma, but putting it this way leads readers to a false presumption: that the bottom line is a fiendishly no-win choice between burning ever-increasing amounts of unhealthy coal, the environmental hazards of which are every day becoming more obvious, or resorting to expensive and hazardous nuclear power, which the public generally abhors. There is another option. Well before another generation of nuclear plants or large coal plants can be built, photovoltaics will be a viable option in the marketplace.

Electricity is by all odds the preferred energy source for our urbanized, industrialized, computerized society as it marches ahead into the post-industrialized, information-centered era. In 1950, 11 percent of the energy used in the United States was electricity. The figure is now 35 percent, and by 2000 it is projected to be nearly 50 percent.[3] Assuring an adequate supply of electricity from sources compatible with our other needs is of prime importance.

As John Naisbitt points out in *Megatrends,* a phenomenon of our day is the rapid growth of new businesses due to entrepreneurs. "We are shifting from a managerial society to an entrepreneurial society" he writes, pointing out that most new jobs are in entrepreneurial firms. Silicon Valley syndrome is but one example of a whole movement away from rigidity in the business-financial-industrial world. Our rapid growth of jobs is the envy of Europe, where high unemployment seems to be an intractable problem in a business environment that is not as open to innovative practices.

Photovoltaics fits perfectly into this setting, both as an activity that is itself the outgrowth of a remarkable entrepreneurial effort and as a decentralized power source for a freed-up, mobile business community. From the Cherry Hill conference of 1973, which kicked it off, to the growth of the maturing photovoltaic industry today, photovoltaics has become more than a technology with great possibilities. It now is ready for use as a fundamental building block in our movement toward a cleaner, healthier, safer, more secure world.

Photovoltaics is pervasive and ubiquitous. It fits the mood of our times and the needs of people everywhere as it comes quietly on the scene.

NOTES

1. As used here, "cost" is synonymous with fair market price, the price at which a willing seller would sell and a willing buyer would buy under stable market conditions. "Cost" thus includes a reasonable profit to the manufacturer.
2. *Washington Post,* 27 April 1984.
3. Eugene Oatman, EPRI.

FURTHER READING

Bush, Leon R. "Photovoltaic Energy Systems Summary Report," Aerospace Corp., Report # DOE-CS/3-622-T1, June 1983.

Energy Sciences-Wholesale Products Catalog ES845, 16728 Oakmont Avenue, Gaithersburg, Md. 20877.

Fueger, Harold B. "The Growing Importance of Electricity: Early Warnings of a Developing Crisis." Paper presented at the 11th Energy Technology Conference, Washington, D.C., March 19, 1984, 11th Energy Technology Conference, 966 Hungerford Dr., #24, Rockville, Md. 20850.

Maycock, P. D., and Stirewalt, E. N. "Experts Answer Key Questions About PV Systems." *Architectural Technology,* Spring 1984.

Naisbitt, John. *Megatrends.* New York: Warner Books, 1984.

Oatman, Eugene N. (EPRI), "If the Lights Go Out?—The Day After." Paper presented at the 11th Energy Technology Conference, Washington, D.C., March 19, 1984, 11th Energy Technology Conference, 966 Hungerford Dr., #24, Rockville, Md. 20850.

Photovoltaics for Residential Applications. Solar Energy Research Institute. SERI/SP-281-2190. Golden, Colo.: Solar Energy Research Institute.

EPILOGUE: WHAT YOU CAN DO

A new technology is unfolding before our eyes—one which could revolutionize our world by augmenting finite energy supplies with limitless energy from the sun. Photovoltaics affords the individual direct access to convenient, versatile, clean electricity. The homeowner can own and control his own energy system.

At the same time, photovoltaics offers unique opportunities for reorienting two other technologies that are central to our civilization—the automobile and the electric utility, both now in dire straits.

It is a dramatic moment when post-industrial civilization is offered a new tool such as this and must choose how to apply it—whether to pick it up and test its possibilities or allow it to languish. The test is whether collectively we have the acumen, the wit, the skill, and daring to put it to good use promptly.

The thought of such change is stupendous and seemingly dwarfs the capability of the individual to have any effect on it. What can the average person do that will help bring about an order that is free from extortion by foreign oil powers, that provides energy without pollution, and that shifts control of energy-producing equipment closer to the people who use the energy? While this goal may appear visionary, there are concrete actions that citizens can take now to assure its realization.

Contacting manufacturers about types and prices of photovoltaic systems for particular needs alerts them to rising public interest. The industry presently sits waiting for clear signs of public responsiveness.

Those who are financially able and willing can purchase and install photovoltaic systems of limited size on their residences. This will help demonstrate the feasibility of photovoltaics and heighten interest in the technology. As small orders come in, industry will be encouraged to expand capacity to meet a growing market. This will maintain the downward trend in prices.

Seriously consider the purchase of an electric auto that can be used frequently for short trips, including commuting to work. In typical urban/suburban use, 80 percent of car trips are less than ten miles long. Purchasing an electric auto will, of course, save gas. More important, it will advance the technology and hasten the day others can follow suit.

Farmers can consider special applications where photovoltaics would work as well as or better than a power line. Electric fences can now be powered by a relatively small photovoltaic array in combination with a battery, and so can a light at a remote location. Such systems seldom need attention. Another frequent application is pumping water from a stream, lake, or well at a point remote from utility power.

Lawyers can investigate local ordinances to see if they restrict the introduction of new equipment such as photovoltaics, and explore ways to remove or ease such impediments. In the spirit of enlightened self-interest, matters such as sun rights, insurability of photovoltaic arrays, and local, state, and federal incentives for solar equipment purchase and installation should be examined.

Insurance companies could prepare now to provide insurance against weather damage, such as hail, to photovoltaic arrays, and other coverage that a homeowner or business person may need. Companies that provide coverage for solar systems should mention it in advertisements to the public.

Local government officials can look into the availability of photovoltaic equipment for special applications. Building and zoning ordinances can be modified so that they will not restrict photovoltaic system introduction; new ordinances can be introduced to positively promote photovoltaic uses.

If smog is a problem, tax incentives for people who buy electric autos would be a positive step. One of the most constructive actions that states and localities could take would be to reduce annual license fees and taxes on electric vehicles. Such measures could be justified on the basis that there is a positive contribution to environmental improvement—a significant benefit to the community in some smog-choked urban areas. In addition, special parking spaces could be equipped with metered electrical outlets where electric vehicles could get recharges during the day. Many states and localities have already exempted solar equipment, including photovoltaics, from taxation, and others might follow their example.

Local governments could enhance the value of future structures by enacting building codes that encourage, if not require, consideration of maximal south-facing roof areas for future addition of photovoltaic arrays.

Municipally owned electric utilities could agree to buy excess power from cooperating homeowners at reasonable, specified rates. If plant expansion is contemplated, officials should look into the rapidly changing economics of using photovoltaics rather than conventional fuels.

Entrepreneurs might consider opening or participating in a business that installs and maintains photovoltaic systems.

Industries can consider applying solar photovoltaics somewhere in their operations. Special tax breaks might apply, and a photovoltaic array could relieve uncertainties about future prices for oil or other fossil fuels.

Bankers can consider setting up financing guidelines for photovoltaic systems and then publicize them in promotional literature.

Tradespeople can set up a photovoltaic service and installation company or add this capability to an existing electrical or roofing operation. People with an engineering background may want to look into a photovoltaic distributorship and repair company.

Persons interested in protecting and improving the environment can promote solar photovoltaics, as well as other solar technologies, as a positive step toward reducing air pollution from central power stations; they can call attention to the advantages of electric vehicles for abating pollution in cities and towns.

Chamber of Commerce officials can boost solar businesses as a new area of commercial activity. They can encourage photovoltaic research and development or manufacturing ventures to locate in their areas.

Auto enthusiasts can promote the electric auto through auto hobby clubs. Sponsoring a slide show or a movie on electric vehicles and disseminating literature on the subject can increase public interest. Information is also available from the Electric Vehicle Development Corp. (20833 Stevens Creek Blvd., Cupertino, CA 95014), which is promoting commercialization, and from the Department of Energy.

Advertising executives might suggest to clients some ideas for tie-in advertising with solar photovoltaics. A photovoltaic array in the background of an advertisement provides a look-ahead tone. A photovoltaic array on the roof or in the yard will soon be something of a status symbol, one that also emphasizes frugality and good sense.

Architects can encourage clients to examine the advantages of solar installations and the financial prudence, as well as future convenience, of providing a large, clear south-facing roof area at optimum slope for the latitude, and perhaps to plan to use portions of south-facing walls as well. People erecting new houses or other structures should be advised that photovoltaic systems will be economically competitive in many parts of the United States within a few years, and already may be preferable to extending a power line a few hundred yards.

Building or electric contractors should acquaint themselves with photovoltaic systems layout, installation, and testing, and the benefits when used in conjunction with a heat pump.

Teachers can make photovoltaics a part of the curriculum. High school science teachers can include simple experiments with photovoltaic cells. Students should be given hands-on experience in measuring voltage and current under different lighting conditions. Physics classes could determine "I-V" curves and compute cell efficiency. Biology teachers can call attention to the advantages of photovoltaics and other renewable energy systems from the standpont of environmental cleanliness. Industrial arts teachers can demonstrate examples of various applications and basic maintenance procedures. College physics instructors can include the principles of photovoltaic conversion and some problems related to basic operational factors. Driver education teachers can include information on electric autos in their courses and may even be able to arrange a demonstration through a dealer or private owner.

As Winston Churchill once observed, "In politics the only thing that counts is numbers." Letters and telegrams to congressmen and senators do contribute to the critical decision-making process of government and are especially needed to urge that tax credits be extended.

Enlightened government policies have nurtured photovoltaic technology for more than two decades, bringing it from a laboratory curiosity to the point that it is ready to be used by society. Government can and will continue to help, and the government role will be critical. But the federal government cannot force this new technology onto an unwilling or indifferent public or deliver it on a silver platter.

The political leadership can at best coax and cajole its constituency toward a new energy future. The powerful forces tugging this way and that are too unruly for the entire nation to turn willingly and in unison in a new direction. The politicians will not propel this country into a realm of renewable energy sources. They can't, they won't, and they shouldn't. But when the citizens are ready, the leadership will respond.

APPENDIX 1
A BRIEF HISTORY OF PHOTOVOLTAICS

Until the advent of the space program, photovoltaic technology remained the province of research scientists. Although the conversion of light into electricity had great impact on inquiries into the nature of matter and the universe, valiant entrepreneurial efforts could not move photovoltaics into the marketplace.

The French scientist Becquerel was the first to observe the photovoltaic phenomenon. In 1839, while experimenting with an electrolytic cell made up of two metal electrodes placed in an electricity-conducting solution, he discovered he could generate more current simply by shining light on the cell. Becquerel further found that the current changed when the wavelength of the light was changed.

Willoughby Smith discovered the photoconductivity of selenium in 1873, and Adams and Day observed the photovoltaic effect in solid selenium three years later. In 1883, Fritts described the first selenium photovoltaic cell, and in 1887, Hertz discovered that ultraviolet light altered the lowest voltage capable of causing a spark to jump between two metal electrodes. In 1904, Hallwachs found that combination copper-cuprous oxide structures were photosensitive.

By 1905, it was known that at any fixed wavelength (frequency), the number of electrons emitted from a substance depended on the intensity of the light bombarding it, and that the maximum kinetic energy (energy of motion) of the electrons varied with the light's wavelength.

In that year, Einstein astounded the world of science by publishing three extraordinary papers. One described the special theory of relativity; a second explained the photoelectric effect. In drawing on contemporary quantum theory, which described the discrete packets of energy occurring at the atomic level, Einstein deduced that the behavior of light could be described in terms of separate particles as well as by the more familiar wave metaphor. He theorized that collisions between electrons and particles of light drove the electrons out of a substance; he precisely formulated the maximum kinetic energy of the emitted electrons as varying by a universal constant with the frequency of the light, after subtracting the amount of work the electrons exerted to reach and escape the surface of the material. In 1916, Millikan provided experimental proof of the theory.

In 1914, the existence of a barrier layer in photovoltaic devices was noted. Much of the subsequent scientific and engineering work exploited the copper-copper oxide interaction, but a perfected selenium device drawing on this developmental effort finally outperformed copper by about a factor of ten. By 1941, selenium devices were converting sunlight to electricity with an efficiency of about 1 percent. In 1951, a new technique called a "grown p-n junction" enabled the production of a single-crystal cell made of germanium.

By diffusing an impurity through silicon crystals, scientific teams including Chapin, Fuller, Pearson, and Prince[1] at AT&T's Bell Laboratories and Rappaport,[2] Loferski, and Jenny at RCA were able to push solar conversion efficiencies to 6 percent in the mid-1950s. While materials such as gallium arsenide and indium phosphide were found to have higher theoretical efficiencies, silicon still yielded the highest experimental results: Factors such as diffusion length and the stability of the material affected final performance. For this reason, and since the extensive prior research in silicon militated against spending large amounts of money to bring the knowledge of alternative materials up to an equal level, silicon continued to be the favored material for use in photovoltaic cells, a position it holds today.

Western Electric began to sell commercial licenses for silicon photovoltaic technology in 1955. Two of the most successful products to emerge were photovoltaic-powered dollar bill change machines and photovoltaic-powered devices that decoded punch cards and tape. Businesses tried to market such products as photovoltaic-powered hearing aids, radios, highway construction warning flashers, and various toys, but satisfactory technical performance could not overcome the economic advantages of conventional power sources.

The photosensor industry soon diverged from solar power generation. In the camera industry, the use of silicon in light meters lost out to selenium because the latter's spectral response better matched that of the human eye.

Although researchers had raised the laboratory performance of solar cells to 14 percent efficiency by 1958, market penetration problems appeared insuperable. According to the late Paul Rappaport, at one point RCA came very close to putting the technology back on the shelf—another example of an interesting idea with no place to go.

No one foresaw Sputnik. The successful use of photovoltaic cells in space carried the technology forward at a rapid pace and provided the impetus for the investigation of photovoltaic applications on earth during the past decade.

---

1 Dr. Morton B. Prince is research director of the federal photovoltaic program in the U.S. Department of Energy.

2 Dr. Paul Rappaport was a pioneer in photovoltaics and the first director of the Solar Energy Research Institute, Golden, Colo.

APPENDIX 2
MAJOR PHOTOVOLTAIC ENERGY PROJECTS IN THE UNITED STATES

1. The following intermediate-size federally funded projects were constructed in 1980–82 under DOE program research and development announcements (PRDAs).

Contractor	*Location*	*Application*	*Size (peak kilowatts)*
Concentrators:			
Acurex	Kauai, HI	Hospital	60*
Arizona Public Service	Phoenix, AZ	Airport	225
BDM	Albuquerque, NM	Office building	47
E-Systems	Dallas, TX	Airport	27
Flat Panels:			
Lea County Electric Cooperative	Lovington, NM	Shopping center	100
New Mexico State	El Paso, TX	Computer at power station	20
Science Applications	Oklahoma City, OK	Science & art center	150
Solar Power Corp.	Beverly, MA	High school	100

*Dismantled in 1983.

2. Other federally supported projects now in operation are:

Contractor	*Location*	*Size (peak kilowatts)*
Concentrators:		
Mississippi County Community College	Blytheville, AR	250
Flat Panels:		
Mt. Laguna Air Force Base	Mt. Laguna, CA	60
Natural Bridges National Monument	Utah	100
Papago Indian village	Schuchuli, AZ	3.5
Agricultural test facility	Mead, NE	25
AM radio station	Bryan, OH	25
Georgetown University International Center	Washington, DC	300

3. Residential Experiment Stations (RES)

Northeast Residential Experiment Station
Hanscom Field
Concord, MA
(Operated by MIT Energy Laboratory, Massachusetts Institute of Technology, Cambridge, MA 02139)

Southeast Residential Experiment Station
Florida Solar Energy Center
300 State Road 401
Cape Canaveral, FL 32920

Southwest Residential Experiment Station
New Mexico Solar Energy Institute
Box 3SOL
Las Cruces, NM 88003
(Adjoins campus of New Mexico State University)

4. Central Station Projects

Contractor	*Location*	*Size*
a. With partial Federal Support:		
Sacramento Municipal Utility District	Sacramento, CA	1 megawatt
b. Private Sector:		
ARCO Solar	Hesperia, CA	1 megawatt
United Energy Corp.	Davis, CA	Greater than 1 megawatt
United Energy Corp.	Borrego Springs, CA	3-5 megawatts
United Energy Corp.	Barstow, CA	Greater than 1 megawatt
ARCO Solar	San Louis Obispo (Carissa Plain), CA	7 megawatts

APPENDIX 3

SOURCES OF ADDITIONAL INFORMATION ABOUT PHOTOVOLTAICS

The principal clearing house for all Department of Energy (DOE) reports is DOE Technical Information Center (TIC), Oak Ridge, Tennessee 37830. The folder TID-4600 R2 describes the functions and services the Center provides. Individual reports can be ordered directly by specifying title and, if possible, author and report number. TIC publishes bibliographies by subject area. Many users might want *Solar Energy: A Bibliography* (TID-3351), which contains references prior to 1976. In addition, *Solar Energy Update,* which indexes all new additions each month, is available from TIC on subscription. Many public libraries maintain collections of DOE reports on microfiche.

The Solar Energy Information Data Bank, operated by the Solar Energy Research Institute, 1617 Cole Boulevard, Golden, Colorado 80401, periodically publishes a *Reading List on Photovoltaics.* This identifies books and articles by private publishers, as well as government documents considered to be useful source materials on photovoltaics.

The primary source for all government reports is the National Technical Information Service, 5285 Port Royal Road, Springfield, Virginia 22181. Principal publications are available from them at nominal charge. NTIS publishes *Abstract Newsletter on Energy* (NTISUB/097), a weekly guide to new materials in the energy field (subscription price: $80 per year). Also available from NTIS are 150,000 foreign technology titles.

Additional government reports are available from the Superintendent of Documents, U.S. Government Printing Office, Washington, D.C. 20402. The Government Printing Office provides periodic lists of new reports to which one may subscribe free of charge.

Information on federal programs and activities in photovoltaics is available from Photovoltaic Energy Technology Division, U.S. Department of Energy, Forrestal Building, Washington, D.C. 20585.

APPENDIX 4

U.S. MANUFACTURERS AND SUPPLIERS OF PHOTOVOLTAIC MODULES, SYSTEMS, AND EQUIPMENT

The following chart is based primarily on data from *Photovoltaic Product Directory and Buyers Guide,* April 1984, PNL-5052, which is available from the National Technical Information Service, 5285 Port Royal Road, Springfield, Virginia 22161. The authors are indebted to Battelle Pacific Northwest Laboratory for the use of this material. The authors and Rodale Press make no warranty, express or implied, and assume no legal liability or responsibility regarding the accuracy of the information or the efficacy or usefulness of any apparatus, product, process, or design.

	PV Modules & Concentrators	Custom-Built PV Systems	Voltage Regulators	Inverters/ Converters	Batteries	PV-Powered Pumps	PV-Powered Battery Chargers	PV-Powered Refrigerators	PV-Powered Vent Fans	Specialty PV Appliances
AAI		●								
Abacus Controls		●		●						
AFG Industries	●									
American Power Conversion				●						
Applied Solar Energy		●								
ARCO Solar	●		●				●		●	
Arctic Cold								●		
Arthur D. Little	●									
A.Y. McDonald		●				●				
Balance of System Specialists			●							
Best Energy Systems				●						
Braden Wire & Metal Products										●
C & D Batteries					●					
Chronar	●									
Crystal Systems	●									
Delco Remy					●					
Dynamote				●						
Energy Materials	●									
Entech	●									
Exide					●					
Free Energy Systems		●					●			
Gates Energy Products					●					
General Electric		●								
Globe-Union		●								
Heart Interface				●						
Helionetics		●		●						
Hydrocap					●					
Intersol Power	●									
Iota Engineering		●								

	PV Modules & Concentrators	Custom-Built PV Systems	Voltage Regulators	Inverters/ Converters	Batteries	PV-Powered Pumps	PV-Powered Battery Chargers	PV-Powered Refrigerators	PV-Powered Vent Fans	Specialty PV Appliances
March Manufacturing		●								
Mobil Solar Energy	●	●								
Norcold								●		
Nova Electric				●						
Opto Technology	●									
Parker McCrory									●	●
Photovoltaic Energy Sys.	●	●								
Polar Products								●		
Polaroid	●									
Power Sonic					●					
Pulstar						●				
Roger Ethier Assoc.		●								
Silicon Materials		●								
Silicon Sensors	●	●	●							
Solar Contractors		●								
Solar Design Assoc.		●								
Solar Usage Now		●								
Solarex	●	●	●				●		●	●
Solarwest Electric		●				●	●			
Solavolt International	●	●								
Solec International	●		●				●			
Solenergy	●	●								
Sollos		●								
Specialty Concepts			●							
Spectrolab		●								
Spire		●								
Surrette Storage Battery					●					
Teledyne Inet		●								
3M	●									

	PV Modules & Concentrators	Custom-Built PV Systems	Voltage Regulators	Inverters/ Converters	Batteries	PV-Powered Pumps	PV-Powered Battery Chargers	PV-Powered Refrigerators	PV-Powered Vent Fans	Specialty PV Appliances
Tideland Signal	●	●					●			
Topaz				●						
TriSolar		●				●				
United Energy	●	●								
Westinghouse Electric	●									
William Lamb									●	
Wilmore Electronics				●						
Windworks		●								
Zomeworks		●								

ADDRESS LIST

AAI Corp.
PO Box 6767
Baltimore, MD 21204
(301) 628-3481

Abacus Controls, Inc.
PO Box 893
Somerville, NJ 08876
(201) 526-6010

AFG Industries, Inc.
PO Box 929
Kingston, TN 37662
(615) 245-0211

American Power Conversion Corp.
89 Cambridge St.
Burlington, MA 01803
(617) 273-1570

Applied Solar Energy Corp.
PO Box 1212
City of Industry, CA 91749
(213) 968-6581

ARCO Solar, Inc.
20554 Plummer St.
Chatsworth, CA 91311
(818) 700-7152

Arctic Cold
3 Old Windsor Rd.
Bloomfield, CT 06002
(203) 242-2211

Arthur D. Little
Acorn Park
Cambridge, MA 02140
(617) 864-5770

A. Y. McDonald
4800 Chavenelle Rd.
PO Box 508
Dubuque, IA 52001
(319) 583-7311, Ext. 227

Balance of System Specialists (BOSS)
7745 E. Redfield Rd.
Scottsdale, AZ 85260
(602) 948-9809

Best Energy Systems
Route 1, PO Box 106
Necedah, WI 54646
(608) 565-7200

Braden Wire & Metal Products, Inc.
PO Box 5087
San Antonio, TX 78201
(512) 734-5189

C & D Batteries
3043 Walton Rd.
Plymouth Meeting, PA 19462
(215) 818-9000, Ext. 305

Chronar Corp.
PO Box 177
Princeton, NJ 08540
(609) 587-8000

Crystal Systems, Inc.
35 Congress St.
Salem, MA 01970
(617) 745-0088

Delco Remy
2401 Columbus Ave.
Anderson, IN 46011
(317) 646-7404

Dynamote Corp.
1200 West Nickerson
Seattle, WA 98119
(206) 282-1000

Energy Materials Corp.
PO Box 353
Ayre Rd.
Harvard, MA 01051
(617) 456-8707

Entech
Dallas-Ft. Worth Airport
Dallas, TX 75261
(214) 272-4705

Exide Corp.
101 Gibraltar Rd.
Horsham, PA 19044
(215) 441-7480

Free Energy Systems, Inc.
Holmes Industrial Park
Holmes, PA 19043
(215) 583-4780

Gates Energy Products
1050 S. Broadway
Denver, CO 80217
(303) 744-4806

General Electric
PO Box 8661, Room 114
Philadelphia, PA 19101
(215) 962-5835

Globe-Union, Inc.
Battery Division
5757 North Green Bay Ave.
Milwaukee, WI 53201
(414) 228-2581

Heart Interface
1626 S. 341st Pl.
Federal Way, WA 98003
(206) 838-4295

Helionetics
Delta Electronic Control Division
17312 Eastman St.
Irvine, CA 92714
(714) 546-4731

Hydrocap Corp.
975 N.W. 95th St.
PO Box 380698
Miami, FL 33138
(305) 696-2504

Intersol Power Corp.
11901 W. Cedar Ave.
Lakewood, CO 80228
(303) 989-8710

Iota Engineering, Inc.
4700 S. Park Ave.
Suite 8
Tucson, AZ 85714
(602) 294-3292

March Manufacturing
1819 Pickwick Ave.
Glenview, IL 60025
(312) 729-5300

Mobil Solar Energy Corp.
16 Hickory Dr.
Waltham, MA 02154
(617) 890-0909

Motorola (see Solavolt)

Norcold Inc.
1510 Michigan St.
Sidney, OH 45365
(513) 492-1111

Nova Electric Manufacturing Co.
263 Hillside Ave.
Nutley, NJ 07110
(201) 661-3434

Opto Technology, Inc.
Solar Systems, Inc.
1674 S. Wolf Rd.
Wheeling, IL 60090
(312) 537-4277

Parker McCrory
Parmak Division
2000 Forest St.
Kansas City, MO 64108
(816) 221-2000

Photovoltaic Energy Systems, Inc.
2401 Childs La.
Alexandria, VA 22308
(703) 780-9236

Polar Products
680 Stone Canyon Rd.
Los Angeles, CA 90077
(213) 476-0082

Polaroid Corp.
730 Main St.
Cambridge, MA 02139
(617) 577-2255

Power Sonic Corp.
PO Box 5242
3100 Spring St.
Redwood City, CA 94063
(415) 364-5001

Pulstar Corp.
Baird Center
619-P S. Main St.
Gainsville, FL 32601
(904) 373-5707

Roger Ethier Assoc.
205 Franklin St.
Alexandria, VA 22314
(703) 683-2657

Shell (see Solavolt)

Silicon Materials, Inc.
999 E. Arques Ave.
Sunnyvale, CA 94086
(408) 737-7100

Silicon Sensors, Inc.
Highway 18 East
Dodgeville, WI 53533
(608) 935-2707

Solar Contractors & Buildings, Inc.
8 Charles Plaza #805
Baltimore, MD 21201
(301) 727-6740

Solar Design Assoc.
PO Box 653
Lincoln, MA 01773
(617) 259-9426

Solar Usage Now, Inc.
Box 306
420 E. Tiffin St.
Bascm, OH 44809
(419) 937-2226

Solarex Corp.
Marketing
1335 Piccard Dr.
Rockville, MD 20850
(301) 948-0202

Solarwest Electric
232 Anacapa St.
Santa Barbara, CA 93101
(805) 963-9667

Solavolt International (Motorola, Shell)
PO Box 2934
Phoenix, AZ 85062
(602) 231-6407

Solec International, Inc.
12533 Chadron St.
Hawthorne, CA 90250
(213) 970-0065

Solenergy Corp.
171 Merrimac St.
Woburn, MA 01801
(617) 938-0563

Sollos, Inc.
1519 Comstock Ave.
Los Angeles, CA 90024
(213) 820-5181

Specialty Concepts
9025 Eton Ave.
Suite D
Canoga Park, CA 91304
(213) 998-5238

Spectrolab, Inc.
12500 Gladstone Ave.
Sylmar, CA 91342
(213) 365-4611

Spire Corp.
Patriots Park
Bedford, MA 01730
(617) 275-6000, Ext. 223

Surrette Storage Battery Co.
PO Box 3027
Salem, MA 01970
(617) 745-4444

Teledyne Inet
2750 W. Lomita Blvd.
Torrance, CA 90509
(213) 325-5040

3M Corp.
3M Center
St. Paul, MN 55144
(612) 733-1110

Tideland Signal Corp.
4310 Directors Row
Box 52430
Houston, TX 77052
(713) 681-6101

Topaz, Inc.
Powermark Division
9192 Topaz Way
San Diego, CA 92123
(619) 279-0831

TriSolar Corp.
10 DeAngelo Dr.
Bedford, MA 01730
(617) 275-1200

United Energy Corp.
420 Lincoln Center Dr.
Foster City, CA 94404
(415) 570-5011

Westinghouse Electric Corp.
PO Box 10864
Pittsburgh, PA 15236
(412) 892-5600

William Lamb Co.
North Hollywood, CA 91601
(213) 980-6248

Wilmore Electronics Co.
PO Box 1329
Hillsborough, NC 27278
(919) 732-9351

Windworks
Box 44A, Route 3
Mukawongo, WI 53149
(414) 363-4088

Zomeworks
PO Box 712
Albuquerque, NM 87103
(505) 242-5354

FURTHER READING

SEIA 1984 Directory and Buyers Guide. Washington, D.C.: Solar Energy Industries Association, 1984.

APPENDIX 5

PROJECTED COSTS OF PHOTOVOLTAIC MODULES AND SYSTEMS

(1984 Dollars per Peak Watt)

The following figures are for large-scale procurements (1 megawatt). Marketplace values for 1984 are included for comparison.

	1984	*1986*	*1990*
Modules	6.00	4.00	1.50
Balance of system:			
Power conditioning	0.50	0.30	0.20
Racks/mounts	1.00	0.50	0.40
Wiring	0.20	0.10	0.10
Installation	1.00	0.50	0.40
Engineering/profit	1.30	0.70	0.50
Installed system	10.00	6.10	3.10

APPENDIX 6
PRINCIPAL CELL MATERIALS AND STRUCTURES

Thin film materials and cell structures that may be used to meet the long-range cost goal of \$0.30–\$0.80 per peak Watt (1984 dollars) for flat-plate collectors.

- **Silicon films**
  Polycrystalline
  Amorphous
- **Non-silicon films**
  Copper sulfide–cadmium sulfide
  Copper sulfide–cadmium zinc sulfide
  Gallium arsenide
  Indium phosphide–cadmium sulfide
  Cadmium indium selenide–cadmium sulfide
  Cadmium telluride
  Cadmium selenide
  Zinc phosphide
  Copper oxide
  Copper ternaries
  Zinc silicon arsenide
  Cadmium silicon arsenide
  Boron arsenide
  Organics
- **Cell structures**
  p-n junctions
  Heterojunctions
  Schottky barriers
  Metal insulator semiconductor (MIS)

Semiconductor insulator semiconductor (SIS)
Luminescent concentrator
Thermal photovoltaic concentrator
Electrochemical cell

Technological approaches for lowering silicon solar cell costs to meet the goal of $1.20–$1.50 per peak Watt for flat-plate collectors.

- **Low-cost polysilicon ($20–$30 per kilogram)**
- **Silicon crystal growing and wafering**
  Advanced Czochralski
  Casting
  Ribbons
- **Simplified cell structures for:**
  p-n junctions
  Heterojunctions
  Schottky barriers
  Metal insulator-semiconductor (MIS) junctions
  Semiconductor-insulator-semiconductor (SIS) junctions

APPENDIX 7
EARTH'S ENERGY BALANCE

This illustration shows the major paths of energy flow into the Earth's system and from the Earth into space. Energy received from the sun must be continuously dissipated into space, or the Earth's temperature would rise until it reached a new equilibrium point. Energy escapes into space mainly through reradiation in the infrared portion of the spectrum (upper right in diagram). However, an increasing level of carbon dioxide in the atmosphere is gradually closing off this escape route. (General concept and basic values courtesy of M. King Hubert, "The Energy Resources of the Earth," *Scientific American*.)

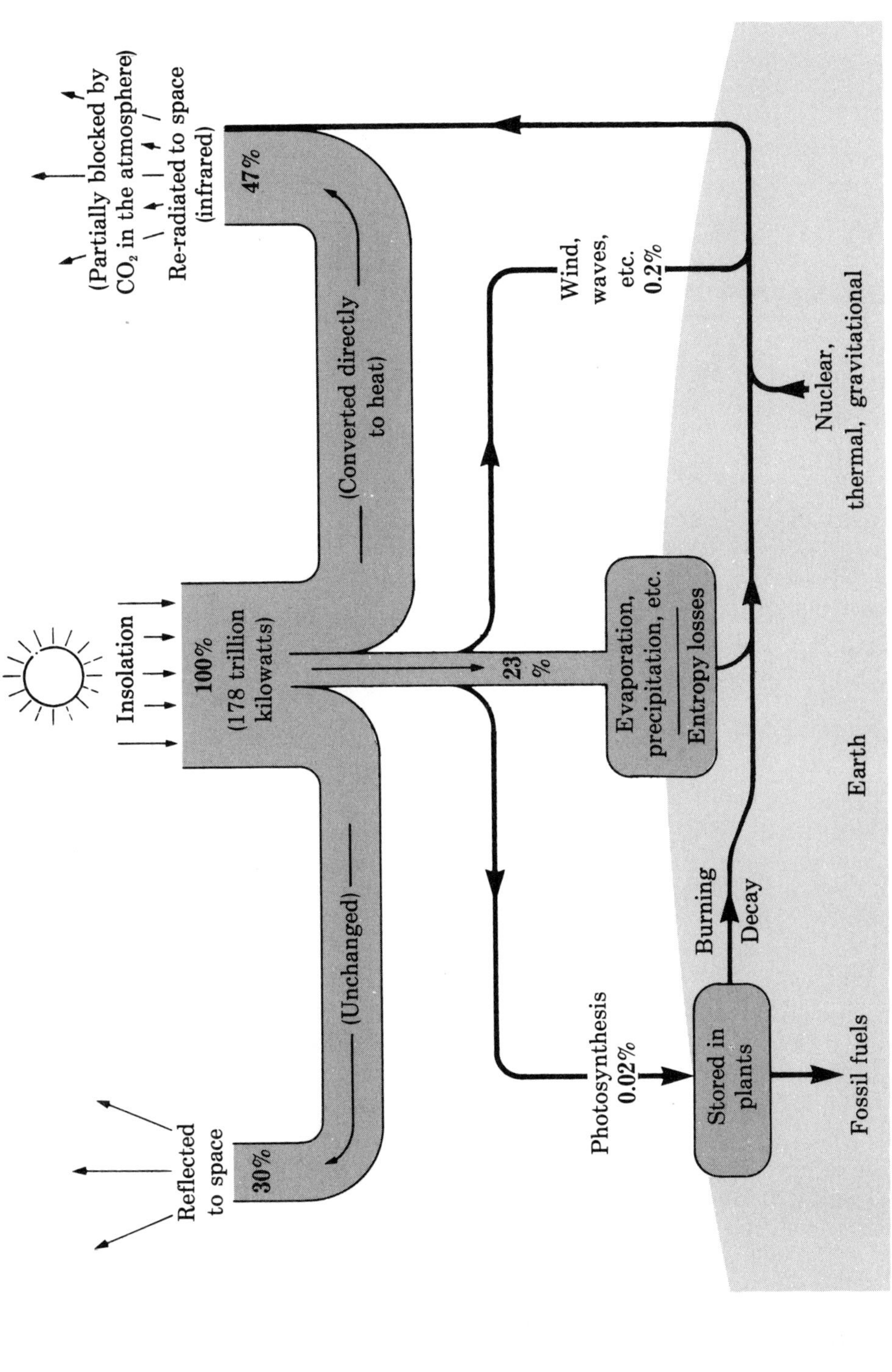
Insolation
100%
(178 trillion
kilowatts)
30%
(Unchanged)
Reflected
to space
47%
(Converted directly
to heat)
(Partially blocked by
CO_2 in the atmosphere)
Re-radiated to space
(infrared)
23
%
Evaporation,
precipitation, etc.
Entropy losses
Wind,
waves,
etc.
0.2%
Photosynthesis
0.02%
Stored in
plants
Burning
Decay
Fossil fuels
Nuclear,
thermal, gravitational
Earth

APPENDIX 8

CURRENT AND PROPOSED U.S. PHOTOVOLTAICS PROJECTS WITH OTHER NATIONS

1. Current and Proposed U.S. Bilateral Agreements in Photovoltaics

Program Title	*Activities*	*Cognizant Organization*	*Approximate Total Budgets ($ Thousands)*
Saudi Solar Village Project (Part of Saudi Arabia-U.S. Joint Program for Cooperation in the Field of Solar Energy)	500-peak-kilowatt village power system; performance evaluation, 3-year operation, and training support	U.S. DOE and Saudi Arabia; managed by Solar Energy Research Institute	20,000
U.S.-Spain Treaty of Friendship and Cooperation	Joint development of two-sided solar cells	U.S. DOE and Spain; supported by Solar Energy Research Institute	1,000
U.S.-Mexico Bilateral Agreement on Energy	Basic photovoltaic research and village power source demonstration	U.S. DOE and Mexico; supported by Solar Energy Research Institute	Not yet determined.

Proposed U.S.-Israel Cooperative Projects on Solar Energy	Proposed research and development on luminescent solar	U.S. DOE and Israel; supported by Solar Energy Research Institute	100
Proposed U.S.-Japan Solar Energy Agreement	Joint research and development in photovoltaics	U.S. DOE and Japan; supported by Solar Energy Research Institute	Not yet determined.

2. Recent International Photovoltaic Projects at the Agency for International Development (AID)

Project Site and Type	*Cognizant Bureau Office within AID and Other Agencies*
Tunisia	
Photovoltaic power for village. A demonstration village project of photovoltaic cells used with power pumping, lighting, and cottage	Near East Bureau with technical advice from NASA Lewis Research Center
Egypt	
Possible inclusion of photovoltaics in a larger solar-wind project	Near East Bureau
Morocco	
Larger renewable energy project, which allows for photovoltaic experimentation among other technologies.	Near East Bureau
Nepal	
Trickle-charge battery system, for use in radio communication. This system is part of a project in teacher training. Radio receivers and cassette tape recorders in remote villages are photovoltaic powered.	Asia Bureau

Bangladesh	
Radio receiver system, powered by photovoltaic cells, as part of a larger project of flood and cyclone warning for rural villages. Approximately 2,200 radio receivers will be distributed (50 Watts each), most of which will be photovoltaic powered.	Asia Bureau in conjunction with Foreign Office of United States Disaster Assistance
Upper Volta	
1.8-peak-kilowatt photovoltaic-powered pumping system for portable water and grain grinding mill in village. Project is part of larger program on women-development-food production, but also explicitly a test and demonstration of a photovoltaic system.	Africa Bureau
Mali	
Pumping powered by photovoltaic cells. Part of larger renewable energy project.	Africa Bureau in coordination with Solar Energy Research Institute; Peace Corps volunteers also involved
Tanzania	
Communication (radio) and perhaps refrigeration for health clinics. This is to be a small demonstration project.	Africa Bureau
Rwanda	
Probably limited photovoltaic component of some kind as part of larger village-level renewable energy project.	Africa Bureau
Niger	
Demonstration of use of photovolatics for irrigation pumping as small part of an institution-building project.	Africa Bureau with Nigerian National Solar Energy Laboratory

Thailand No photovoltaic installations. As part of a larger energy project, there is an agreement to keep Thailand informed of the state of the art in photovoltaics.	Asia Bureau
Philippines The pump irrigation component of a larger nonconventional energy project is photovoltaic powered.	Asia Bureau in coordination with Department of Energy
Asia General Asia conference on renewable energy held in Manila, November 1979, included information on photovoltaics.	Asia Bureau
Panama Small photovoltaic component of large energy project. Photovoltaic cells will be used to power radio transceivers and repeaters for remote schools.	Latin America Bureau
Guyana Possible use of photovoltaics with health clinic equipment.	Latin America Bureau
Coordinated Applications Program Research and development and on-site emphasis on photovoltaic-powered refrigeration for health clinics. Not tied to specific sites; siting to be done later, in coordination with in-country missions.	Energy Office and Participating Agency Service Agreement (PASA) with NASA Lewis Research Center

APPENDIX 9

ESTIMATING PHOTOVOLTAIC SYSTEM REQUIREMENTS FOR A RESIDENCE

A general approximation of system size can be made as follows: One peak kilowatt (kW_p) of photovoltaic system capacity in an area of moderate sunshine produces about 2,000 kilowatt hours (kWh) of electricity in one year (2,500 kilowatt hours in Arizona-New Mexico, 1,600 kilowatt hours in New England and the extreme Northwest). This much electricity would be produced by a permanently mounted array sloped about 45° in a generally southward direction. For example, a person who lives in a region of moderate sunshine and uses 6,000 kilowatt hours in a year (actual amount can be determined from monthly electric bills) would need a 3-kilowatt capacity photovoltaic system (6,000 kilowatt hours divided by 2,000).

The following simple calculations enable one to estimate more precisely the size of a photovoltaic system required for a residence and the cost of power from it.

1. From your electric bill for a month, note the number of kilowatt hours used. Divide by 30 to obtain your approximate daily usage (load). Since there is some seasonal variation, it would be more accurate to add the amounts for a year and divide by 365, if you have saved your old bills.

2. On the insolation map (Fig. A9-1) find your approximate location and note the solar energy received daily in that zone. The figures represent approximate kilowatt hours per square meter falling daily on a 45° sloping array, facing south, as a yearly average. (For a stand-alone residence, that is, without utility backup, use Figure A9-2.)

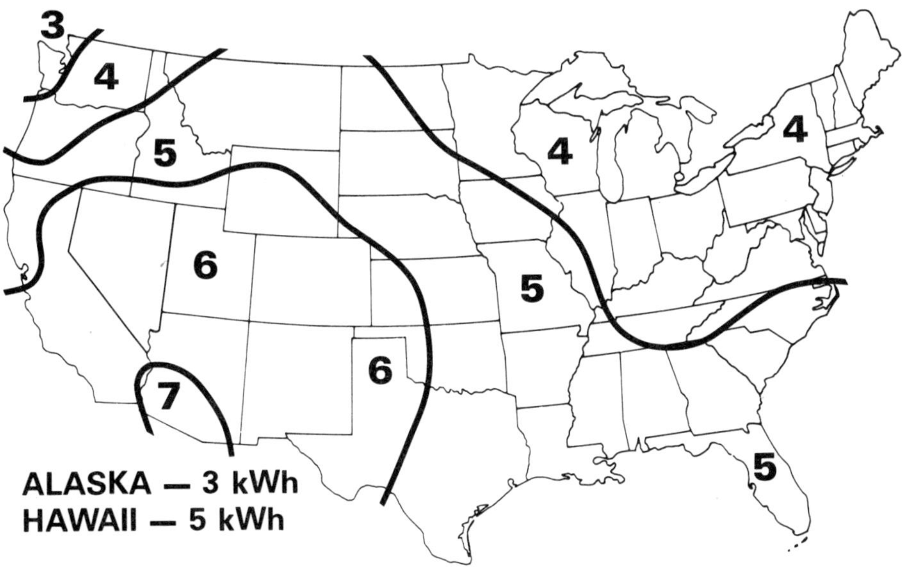

Figure A9-1 Approximate daily insolation falling on a south-facing surface of 1 square meter tilted at an angle equal to local latitude, for an average day in the year (kilowatt hours per square meter). Numerals apply to areas between the isometric curves. A 10° variation in array tilt is not significant. This chart is useful for computing array size to provide a given amount of power on an annual basis. Zones interpolated from a recent study by SERI (*Solar Radiation Energy Resource Atlas of the U.S.*, SERI/SP-642-1037, 1981).

3. Calculate the photovoltaic array area in square meters that you would need to satisfy your average daily electric requirement:

$$\frac{\text{Average daily load requirement (kWh/day)}}{\text{Insolation (kWh/sq meter/day)} \times \text{PV system efficiency}}$$

$$= \text{Array area (sq meters)}$$

Example: Assume you use 17 kWh/day, insolation in your area is 5 kWh/sq meter/day, and the photovoltaic system will be 15 percent efficient.

$$\frac{17 \text{ kWh/day}}{5 \text{ kWh/sq meter/day} \times 0.15} = 22.67 \text{ sq. meters}$$

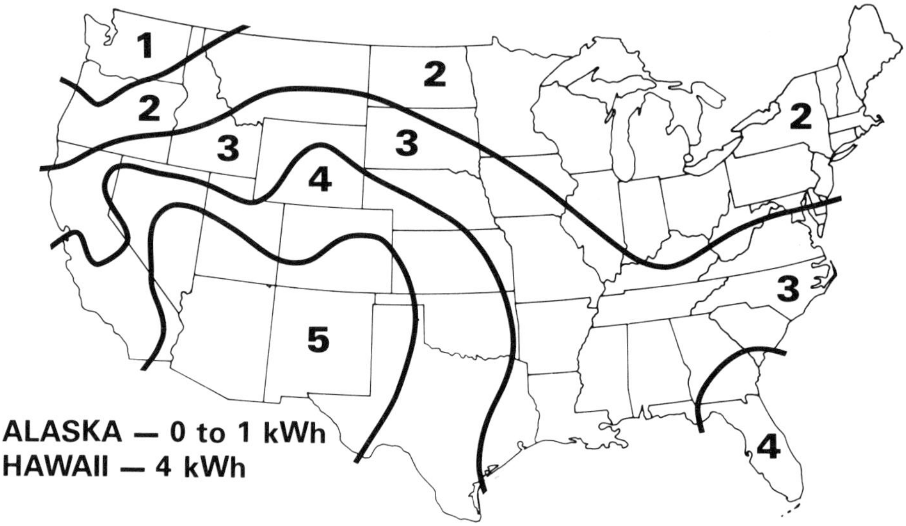

Figure A9-2 Average daily insolation for December received on a south-sloping surface of 1 square meter, tilt equal to latitude (kilowatt hours per square meter). Numerals apply to zones between the curves. Local conditions can produce significant variations from these values, especially in mountainous regions.

One square meter is 10.76 sq. ft. Therefore, 22.67 sq. meters equals 243.9 sq. ft.

4. Calculate the amount of photovoltaic power needed:

$$\text{Array area (sq meters)} \times \text{System Efficiency}$$
$$\times \text{ Solar constant (kW/sq meter)} = \text{Peak power } (kW_p)$$

The "solar constant" is a measure of the sun's energy received on a clear day at high noon at the earth's surface. On a surface facing the sun, it is about 1 kW/sq. meter. Assume system efficiency is 15 percent.

Example:

$$22.67 \text{ sq meters} \times 0.15 \times 1 \text{ kW/sq meter} = 3.4 \text{ } kW_p$$

5. Cost can be estimated directly. Assume that installed system price is $3,000/kW_p ($3/W_p) of generating capacity.

$$\$3{,}000/\mathrm{kW_p} \times 3.4\ \mathrm{kW_p} = \$10{,}200$$

If solar tax credits are available, your final costs are reduced proportionally. For example, a 40% tax credit on the first $10,000 would cut your actual costs by $4,000—from $10,200 to $6,200.

6. To estimate the cost of electricity, assume a system life of 20 years.

$$\frac{\$6{,}200}{17\ \mathrm{kWh/day} \times 365\ \mathrm{days/yr} \times 20\ \mathrm{yrs}} = \$0.05/\mathrm{kWh}$$

Factors not considered in this estimate include the worth of the $6,200 had it been otherwise invested, or the interest paid on it if it were borowed; the higher prices that the homeowner may expect to pay for electricity purchased from a utility in these future years; and maintenance costs on the photovoltaic system.

Tilt Angle (Slope) of Arrays

On a year-around basis, the optimum tilt angle (slope southward) for a fixed array is obtained by adding 10 degrees to the local latitude. In New Orleans at 30°N, optimum slope is 40°; in Minneapolis at 45°N, a 55° slope is best. Exact slope, however, is not critical. For example, a deviation of 15° from optimum reduces efficiency only 3½ percent, and at a 20° deviation the loss is only 6 percent (the loss in efficiency increases as the cosine of the angle). For flat-plate systems, which pick up light coming from all directions, the optimum tilt angle is less important than for concentrator systems, which must track the sun. If the rotation is on one axis only, the tilt is adjusted manually at least every two or three weeks. Two-axis trackers have sensors so that the array follows the sun continually.

APPENDIX 10
THE ELECTRIC AUTO

In the early days of the automobile, electric vehicles greatly outnumbered and significantly outperformed cars with internal combustion engines, but well before World War I, their popularity slacked, and in the 1920s, electric vehicles virtually disappeared. The 1970s oil shortage brought renewed interest in electric vehicles. In 1976, Congress passed the Electric and Hybrid Vehicle Research, Development, and Demonstration Act (Public Law 94-413), which as amended, directs the Department of Energy (DOE) to develop and put into operation for demonstration purposes 7,500–10,000 electric and hybrid vehicles (a hybrid vehicle is one that has a small gasoline or diesel engine to extend the range of its electric propulsion system.)

A typical energy flow diagram for an electric vehicle (Fig. A10-1) shows over half of the energy reaching the wheels, a vast improvement over the internal combustion engine, which wastes at least two-thirds of the fuel within the engine itself, as heat.

Senator Jim McClure, chairman of the Energy and Natural Resources Committee, has said:

> It's important to our national interest to develop a commercial electric vehicle because it will help reduce our consumption of petroleum. Currently, passenger cars and light trucks account for more than 39 percent of our total petroleum consumption. If just 10 percent of the transportation sector were electric cars, it would save 663,000 barrels of oil every day.

Although he acknowledges that much remains to be done, Senator McClure sees solid progress in the last five years:

> In 1977, the best electrical car only reached a top speed of about 50 miles per hour, and could travel just 40 miles in urban traffic before it had to be recharged. In addition, its batteries had to be replaced in less than a year. Today, prototype electric vehicles easily can reach speeds

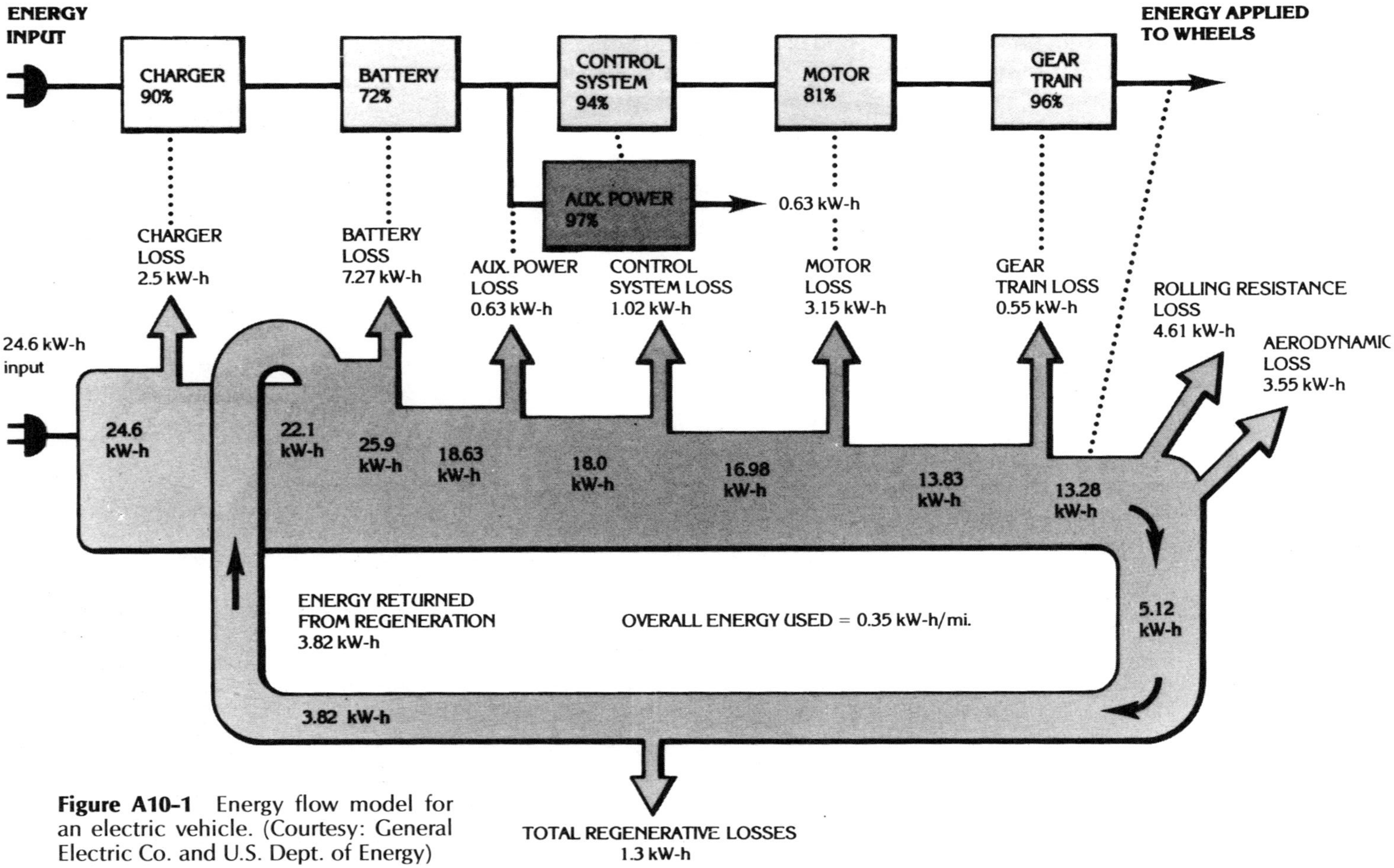

Figure A10–1 Energy flow model for an electric vehicle. (Courtesy: General Electric Co. and U.S. Dept. of Energy)

in excess of 65 miles per hour, and have a range of 93 miles in urban traffic. Their batteries now last as long as two years.

Senator McClure is optimistic about the future:

> While further reduction in battery weight and cost are still necessary, new designs in batteries—such as Gel Cell lead-acid, zinc/bromide, nickel/iron and tubular-plate lead-acid battery systems—offer great promise. Great strides have also been made in the development of electrical power trains. Modern systems now use alternating current instead of direct current. Consequently, motors are more efficient, smaller in size, and lighter in weight.

Two prototype electric test vehicles (ETV) were developed and delivered to DOE in 1979 for road testing. General Electric and Chrysler teamed to produce ETV-1 and Garret Corp. provided ETV-2. Both are 4-passenger subcompacts powered by heavy duty lead-acid batteries. Both have snappy acceleration that approaches present subcompact car performance.

Using the first low-cost, high-power transistor, ETV-1 has optimized all-electric regenerative braking for extended range; microprocessor-controlled electrical functions; abrasion-resistant plastic glazing; charge indicator, which replicates fuel gauge function; structure and battery retention designed to ensure crashworthiness; drag reduced through wind-tunnel tests (yaw effects included); integrated electronics (field chopper and on-board charger); and electric interlocks for safe operation and reduced shock hazard.

The first application of composite flywheel technology to passenger cars, ETV-2 has optimized regenerative braking using flywheel energy storage; flywheel load-levels battery power; microprocessor control; tubular plate applied to golf-cart-size battery; abrasion-resistant plastic glazing; charge and flywheel indicators to replicate fuel gauge function; fiberglass-reinforced plastic in unitized body design; structure and battery retention designed to ensure crashworthiness; electric interlocks for safe operation and reduced shock hazard; and flywheel vacuum maintained by molecular pump.

The Department of Energy continues to support research, development, and testing of vehicles and advanced batteries at a level of more than $13 million a year. The Eaton Corporation is developing both advanced DC and AC propulsion systems. The AC system has a 25-horsepower motor (45 horsepower for two minutes), three-phase induction motor, transistorized inverter, two-speed automatic transaxle, braking regeneration, and an on-board 110-220-Volt battery charger. The propulsion system weighs only 297 pounds.

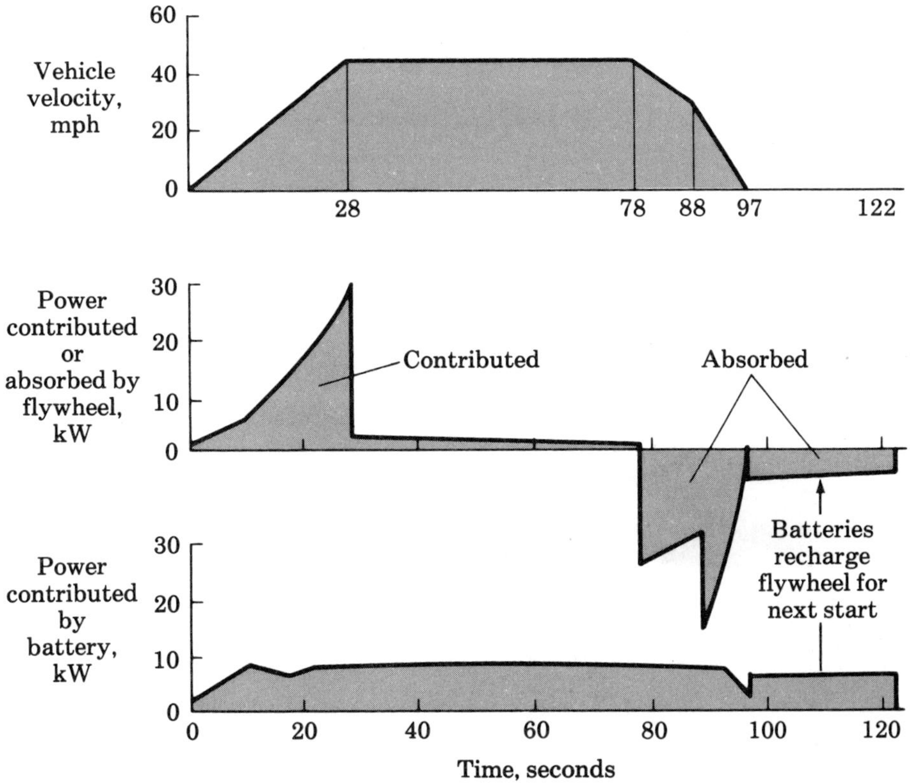

Figure A10-2 Flywheel delivers peak power during acceleration and absorbs braking energy; the battery power remains nearly constant. (Courtesy: Garrett Corp. and U.S. Dept. of Energy)

In a similar cooperative venture with Ford Motor Company, DOE is promoting a parallel AC system (General Electric is a subcontractor).

The early market for electric vehicles will probably focus on electric vans for short-haul work, especially in urban-suburban areas. This will likely be followed by penetration of the private automobile market as vehicle performance improves.

The electric utilities would like to see increased use of electric vehicles because it aids them with "load leveling." The electric vehicle fleet would be recharged largely at night when the utilities have unused base load capacity, their most inexpensive source of electricity. If the utilities can put this capacity to work, the overall efficiency of their operations will be improved.

EPRI, the research and development arm of the electric utilities, recently sponsored—in conjunction with DOE, the utilities, and trade organizations

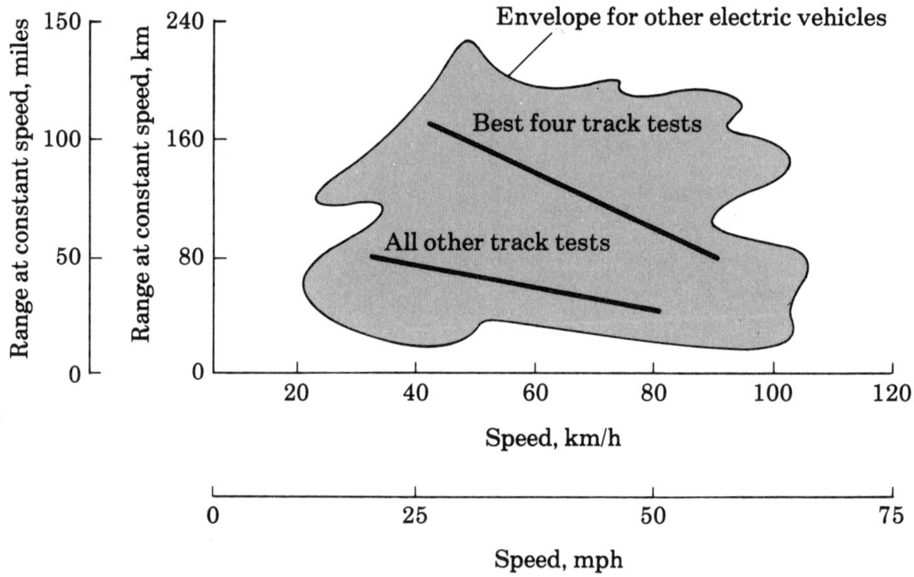

Figure A10-3 Electric vehicle range as a function of speed. The distance an electric vehicle can travel between recharges falls off linearly with increasing speed. Data from 1976-77. (Courtesy: U.S. Dept. of Energy)

—the formation of the Electric Vehicle Development Corporation (EVDC). EVDC's purpose is to promote the commercialization of electric vehicles. Floyd Culler, President of EPRI, says:

> Formation of a national group to focus and consolidate electric vehicle interest and action certainly seems necessary if the goal of commercializing electric vehicles is to be achieved in this country. Further, in view of the potential benefits of the electric vehicle to electric utilities and given our industry's proven ability to plan and implement major end-use load strategies, it is very appropriate for a utility group to take the initiative in creating a national focus.

EVDC plans to promote the testing and demonstration of hundreds of electric vehicles in coming years.

A novel concept developed by an entrepreneur, John S. Reuyl, would provide a hybrid electric automobile in combination with a photovoltaic-powered house. Under normal operation, the photovoltaic system would produce electricity for both house and car. However, the electric car would

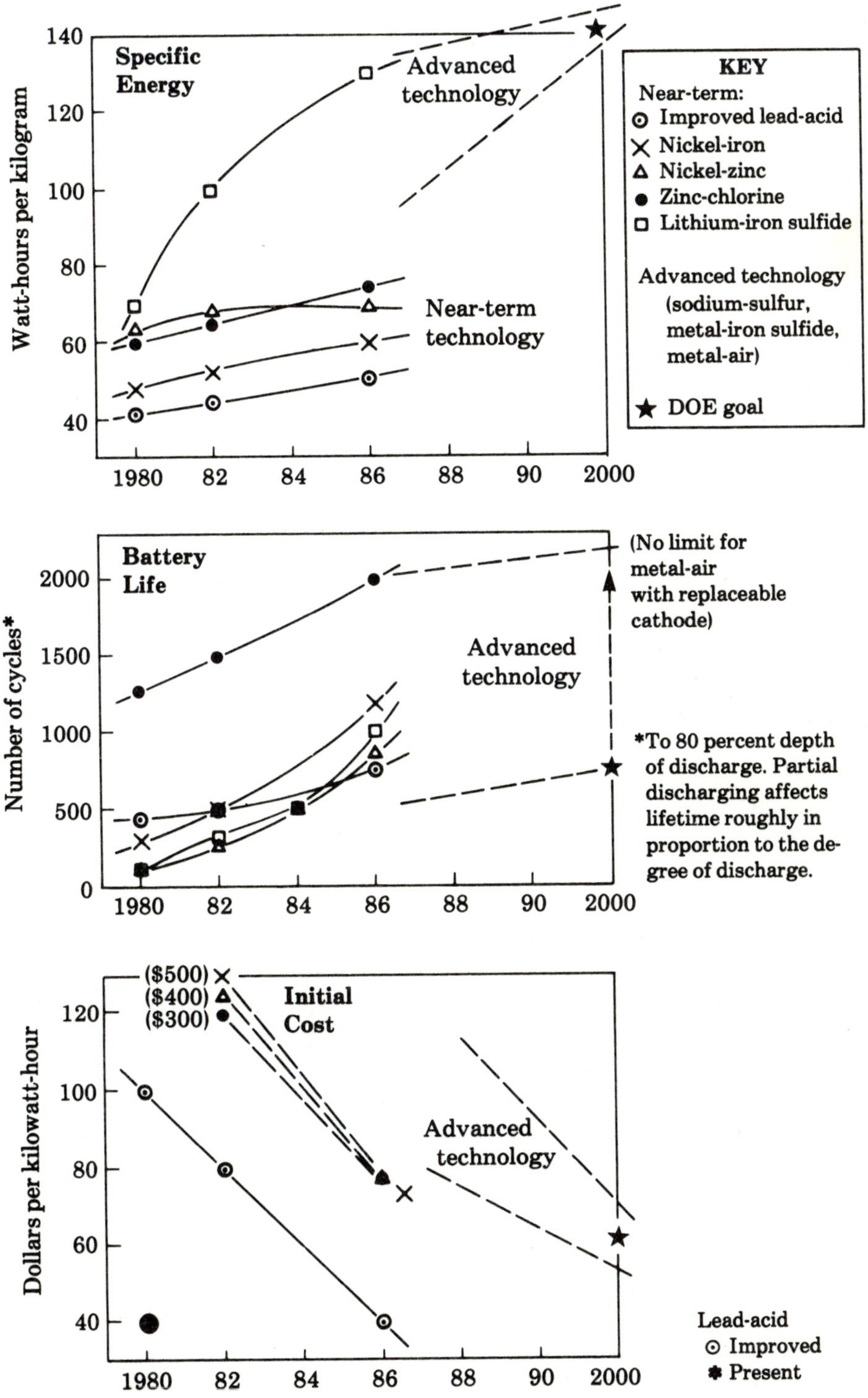

Figure A10-4 Projected vehicle battery characteristics.

also contain a separate, small alcohol engine (alcohol is a renewable energy fuel, although other fuels could be used). During cloudy weather the alcohol engine would be operated at night as a backup unit to recharge the batteries in both the car and the house. In addition, it would give the car increased range.

Significant development efforts are also under way in other countries, notably the United Kingdom, Denmark, West Germany, France, Italy, and Japan.

Senator McClure concludes: "The successful commercialization of electric cars will offer American consumers many advantages over conventional automobiles, including lower operating cost, less chemical and noise pollution, and most importantly, an alternative to the use of petroleum fuels in transportation."

Photovoltaics offers an avenue to automotive transportation free of *all* environmental pollution or hazards from conventional energy sources.

APPENDIX 11

U.S. ENERGY SOURCES AND USES: A COMPARISON BETWEEN 1980 AND PROJECTIONS FOR 1990

Projections for 1990 (see charts following) show dramatic increases anticipated in energy derived from nuclear sources and coal to provide more electricity. Unfortunate consequences will be that most of the additional energy generated will be lost in conversion, and transmission and pollution problems will be intensified.

U.S. ENERGY SOURCES AND USES IN 1980
(IN MILLIONS OF BARRELS PER DAY OF OIL EQUIVALENT — MMBDOE)

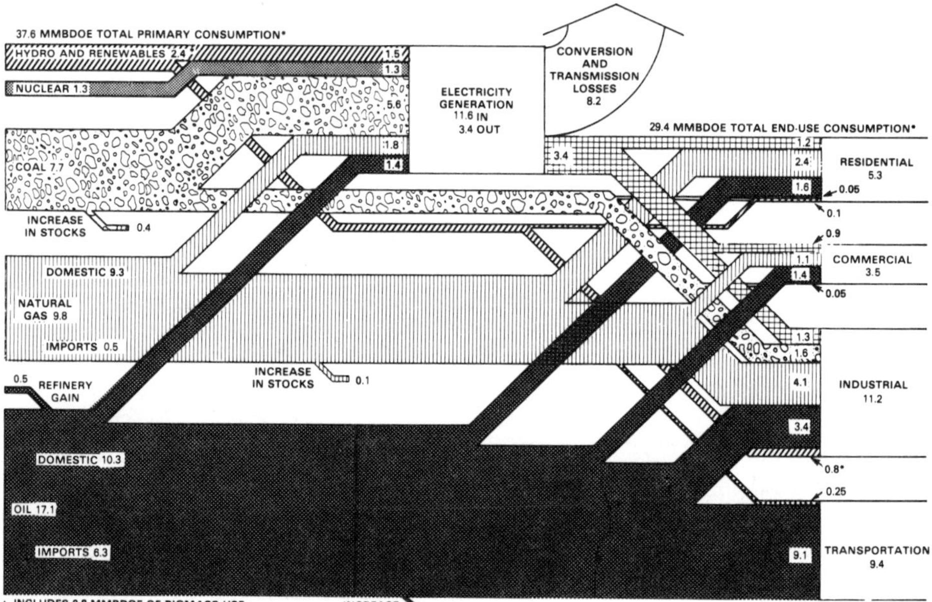

* INCLUDES 0.8 MMBDOE OF BIOMASS USE NOT CURRENTLY ACCOUNTED FOR IN DOE STATISTICS.

SOURCE: PRELIMINARY ESTIMATES; BASED ON ACTUAL DATA THROUGH SEPTEMBER 1980 FROM EIA MONTHLY ENERGY REVIEW AND PROJECTIONS FOR THE 4th QUARTER.

PROJECTED U.S. ENERGY SOURCES AND USES IN 1990
(IN MILLIONS OF BARRELS PER DAY OF OIL EQUIVALENT – MMBDOE)

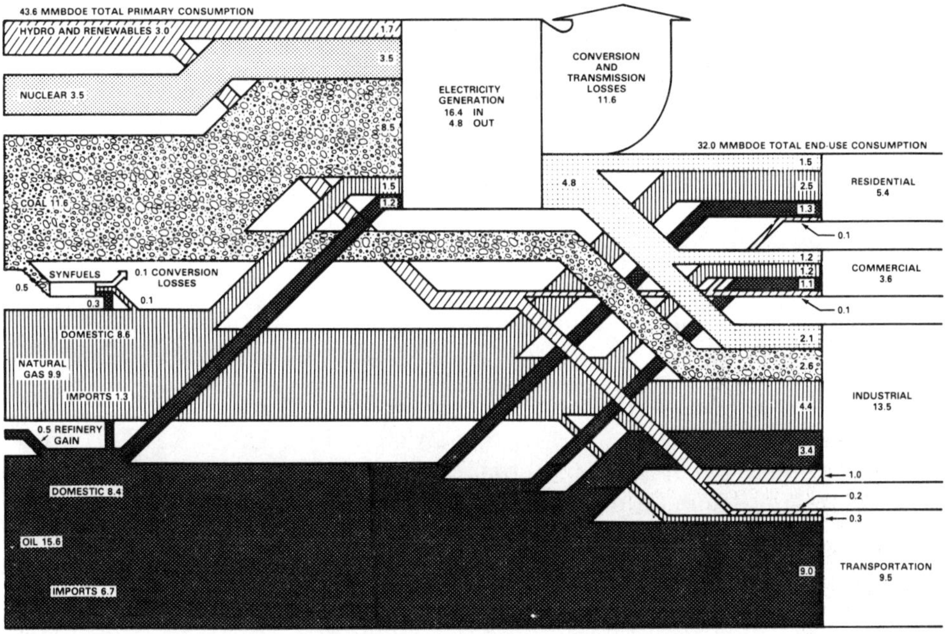

APPENDIX 12
KEY LEGISLATION

Solar Photovoltaic Energy Research, Development, and Demonstration Act of 1978 (Public Law 95-590)

This is the principal law, and the only one, that pertains solely to photovoltaics. It prescribes an aggressive research and demonstration effort over the ensuing decade to double production of photovoltaic cells each year, reaching a cumulative production by 1988 of 4 million kilowatts (4 gigawatts), and to reduce the average installed cost to $1 per Watt. This law authorizes spending $1.5 billion to achieve these objectives, subject to annual appropriations. Among other things, the Act prescribes specific studies to be undertaken, establishes a high-level advisory committee, and encourages a strong international effort.

Department of Energy Act of 1978 (Public Law 95-238)

Section 208, *Civilian Applications,* known as the Tsongas amendment, promotes the use of photovoltaic systems at federal installations and calls for studies of government support for the U.S. photovoltaics industry and of applications in foreign countries.

Crude Oil Windfall Profits Tax Act of 1980 (Public Law 96-223)

One provision qualifies residential purchasers of solar electric systems, which includes photovoltaics, for a 40 percent tax credit on the first $10,000 of the purchase price (limit $4,000). Industrial purchasers may receive a 25 percent tax credit. Scheduled to lapse at the end of 1985.

Public Utility Regulatory Policies Act of 1978 (Public Law 95-617)

Known as PURPA, this act is designed, among other things, to encourage the homeowner and other small users to generate part or all of the electric power they require. To this end, it requires that utilities supply supporting or backup power to these "cogenerators," charge reasonable rates for this, and buy back any excess power produced by cogenerators, at reasonable rates based on the "avoided costs" to the utility.

Federal Photovoltaic Utilization Act (Public Law 95-619—Part 4)

This portion of the National Energy Conservation Policy Act, enacted in 1978, was the basis for the federal program to procure photovoltaic systems for government use, one means for building up the U.S. industry and accelerating experience with the new technology.

In addition to this federal legislation, a number of the states have provided tax incentives to encourage the use of solar energy systems, including photovoltaics. Amounts vary widely. In some states, the tax credit is as high as 55 percent.

The following states have enacted tax credits for residential installations: Alabama, Arizona, California, Colorado, Delaware, Georgia, Hawaii, Idaho, Indiana, Kansas, Kentucky, Massachusetts, Michigan, Minnesota, Montana, Nebraska, New Mexico, New York, North Carolina, North Dakota, Ohio, Oklahoma, Oregon, Rhode Island, Utah, Vermont, Virginia, and Wisconsin.

In addition, these states have enacted tax credits for commercial installations: Arizona, California, Colorado, Georgia, Hawaii, Indiana, Kansas, Kentucky, Nebraska, New Mexico, North Carolina, North Dakota, Ohio, Oklahoma, Oregon, Rhode Island, Utah, Virginia, and Wisconsin.

Note that the credits in some states have impending expiration dates. Check with the energy office in your state.

APPENDIX 13
MISCELLANEOUS CONVERSION FACTORS AND APPROXIMATIONS

1 bbl = 42 gal
1 bbl of crude oil yields 33 gals of gasoline
1 bbl of crude oil is used in propelling a compact car 1,000 miles (at 30 mi/gal)
1 cal = 4.187 Watt-sec
50 GW_p of photovoltaic capacity replaces in one year 1 Q of fossil fuel consumed to make electricity.
50 GW_p provides in one year 100 billion kWh, or 10^5 GWh, of electricity.
1 horsepower = 0.75 kilowatts
1 kWh/sq meter = 86 Langleys
1 Langley = 1 cal(gm)/sq cm
1 Langley = 0.01163 kWh/sq meter
1 megajoule/sq meter = 88.1 Btu/sq ft
1 Q (quad) = 10^{15} Btu
1 Q = 180 million bbls of oil
1 Q = 40 million tons of coal
1 Q = 1 trillion cu ft of natural gas
1 mill bbls oil/day for one year = 2.02 Q
1 sq meter = 10.76 sq ft
1 Watt = 1 joule/sec
1 GWyr = 8,760 GWh = 8.76 billion kWh
1 kWh = 3.6 megajoules
1 kWh = 3,413 BTU total conversion
10,000 BTU at boiler produces 1 kWh

APPENDIX 14

PHOTOVOLTAIC FIRMS AND PROJECTS IN OTHER LANDS

1. European Industrial Firms and Institutions*

Country	*Company*	*Technology Activity*	*Affiliation*
Belgium	Belgosolar (90% government owned)	Cells/Modules	—
	E.N.E. Energie Nouvelle et Envirounement	Cells/Modules	Solarex
England (UK)	Lucas—BP Solar System	Modules/Systems	—
	Solapax	Systems (Import modules from U.S.)	Solarex
France	Societe Francaise de Photopiles	Modules	Photowatt
	Photon	Cells/Modules	Solarex
	Total Oil	Systems	Photon Power
Germany	Wacker Chemitronics	Polycrystalline silicon refining	No
	AEG Telefunken	Modules	No
	Siemens	Modules	No
	NUKEM	Cadmium—Sulfide thin-film cell	—

Country	*Company*	*Technology Activity*	*Affiliation*
Italy	Ansaldo (100% government owned)	Cells/Modules	—
	ENI (75% government owned)		
	a) Solaris	a) Solar Grade Silicon Arc Furnace b) Ingot Casting SEMIX c) Cells/Modules d) Systems	Solarex
	b) Pragma	Solar Grade Silicon	Solarex
	ENEA (government agency)	Research & Experimental Systems	—
	Adriatica Componerti Electtronici	a) Modules b) Solar Grade Silicon	Siemens
Netherlands	Holec Solar Energy	System	Solarex
Switzerland	—	Fully-Integrated PV Production Plant	ENI (Italy) Standard Oil (Solarex)
	Photonetics	Modules	Societe Roma d' Electrict Parsan, Inc.

*Information based on testimony by Dr. R. R. Ferber of JPL to U.S. House of Representatives Subcommittees on Science and Technology (June 1982).

2. European Economic Community Photovoltaic Pilot Projects

kW_p	*Location*	*Application*	*Grid-Conn.*	*Stand-Alone*	*Approximate Cost (Thousand ECU*)*
50	Mont Bouquet (France)	TV transmitter	X		65

kW_p	*Location*	*Application*	*Grid-Conn.*	*Stand-Alone*	*Approximate Cost (Thousand ECU*)*
50	Fota Island, Cork (Ireland)	Dairy farm	X		40
50	Aghia Roumeli, Crete (Greece)	Isolated village		X (Diesel)	38
35	Kaw (French Guyana)	Isolated village		X (Diesel)	36
300	Pellworm Island (W. Germany)	Vacation center	X		2,048
100	Kythnos Island (Greece)	Island power		X (Diesel, wind)	1,985
80	Vulcano Island (Italy)	Island power			1,618
30	Marchwood, Southhampton (U.K.)	Utility augmentation (rural electrification)	X		1,303
65	Tremiti Islands (Italy)	Desalination (reverse osmosis)		X	1,710
63	Chevetogne (Belgium)	Swimming pool (lights & pump)	X		1,219
50	Nice Airport (France)	Air traffic control	X		1,020
50	Terschelling Island (Neth.)	School	X		1,136
45	Giglio Island (Italy)	Water disinfection & food cold storage	X (Diesel)		1,621
44	Rondulinu, Corsica (France)	Village power		X	949
30	Hoboken (Belgium)	Hydrogen production & water pumping	X		621
7–10	Adrano (Italy)	Several small residential systems	X		(?)

*European currency unit. 1 ECU = $0.73 U.S. (October 1984)

3. Japanese Industrial Firms and Institutions

Electro Technical Laboratories
1-1-4, Umerzono, Sakura-mura, Shinji-gun
Ibaraki-ken, Japan

Fuji Electric
Corporate Research and Development, Ltd.
2-2-1, Nagasaka,
Yokosuka-shi
240-01 Japan

Hitachi Ltd.
Nippon Building
No. 6-2, 2-Chome, Ohtemachi
Chiyoda-ku,
Tokyo 100, Japan

Hoxan Corporation
Hoxan Research Laboratories
5 Jo, 2-Chome,
Kikusui, Shitoishi-ku,
Sapporo 003, Japan

Institute of Space and Astronautical Science
6-1, Komaba, 4-Chome, Meguro-ku
Tokyo 153, Japan

Japan Solar Energy Company, Ltd.
11-17 Koga-honmachi,
Fushimi-ku
Kyoto 612, Japan

Kyoto Ceramic Co., Ltd. (Kyocera)
52-11 Inoue-cho, Higashino
Yamashina-ku
Kyoto 607, Japan

Ministry of International Trade and Industry (MITI)
Sunshine Project Headquarters
1-3-1 Kasumigaseki, Chiyoda-ku
Tokyo 100, Japan

New Energy Development Organization (NEDO)
Suhshine Building
3-1-1, Higashi Ikebukuro,
Toshima-ku
Tokyo, Japan

Nippon Electric Co. (NEC), Ltd.
1753, Shimonumabe,
Nakahara-ku
Kawasaski-shi
Kanagawa 211, Japan

Osaka University
Machikaneyama, Toyonaka-shi
Osaka, Japan
Dr. Yoshihiro Hamakawa, Electrical Engineering Department

Sanyo Electric Co., Ltd.
1-18-13, Hashiridani,
Hirakata-shi
Osaka, Japan

Sharp Corporation
Solar Systems Group
Shinjo-cho
Nara 639-21, Japan

Toshiba Corporation
1 Komukai, Toshiba-cho
Saiwai-Ku
Kawasaki 210, Japan

Tokyo Denki Co., Ltd.
1-14-10, Uchikanda, Chiyoda-ku
Tokyo 101, Japan

Tokyo Institute of Technology
12-1, 2-Chome, Ohkayama
Meguro-ku
Tokyo 152, Japan

GLOSSARY

acceptor A dopant material such as boron which has fewer outer shell electrons than required in an otherwise balanced crystal structure, providing a hole, which can accept a free electron.

alternating current (AC) Electric current in which the direction of flow is reversed at frequent intervals, 120 times per second (60 cycles per second), as used in commercial grid power in the United States. Opposite of direct current (DC).

amorphous The condition of a solid in which the atoms are not arranged in an orderly pattern; not crystalline.

ampere, amp A measure of electric current; the flow of electrons. One amp is 1 coulomb (6.3×10^{18} electrons) passing in one second. One amp is produced by an electric force of 1 volt acting across a resistance of 1 ohm.

angstrom A measure of length equal to 10^{-8} cm.

array See *photovoltaic array.*

balance of system (BOS) The parts of a photovoltaic system other than the array: switches, controls, meters, power-conditioning equipment, supporting structure for the array, and storage components, if any. The cost of land is sometimes included when comparing total system costs with the cost of other energy sources.

band gap energy (Eg) The amount of energy (in electron volts) required to free an outer shell electron from its orbit about the nucleus to a free state and thus to promote it from the valence level to the conduction level.

barrier See *cell barrier.*

barrier energy The energy given up by an electron in penetrating the cell barrier; a measure of the electrostatic potential of the barrier.

base load The minimum amount of electric power that a utility must supply in a 24-hour period. Utilities typically operate their most efficient generators (usually their newest and largest) to meet base load demand. *See load; peak load.*

boron (B) A chemical element, atomic number 5, semi-metallic in nature, used as a dopant to make p-silicon.

break-even cost The cost of a photovoltaic system (in dollars per kilowatt of generating capacity) at which the cost of the electricity it produces exactly equals the price of electricity from a competing source.

British thermal unit (Btu) Amount of heat required to raise the temperature of 1 pound of water by 1 degree Fahrenheit.

cadmium (Cd) A chemical element, atomic number 48, used in making certain types of solar cells.

capacity The amount of electricity a system is capable of producing. If a system has a capacity of 5 kilowatts, for example, and if it will produce 2,500 kilowatt hours of electricity for each kilowatt of capacity, then its total annual electric output will be 12,500 kilowatt hours (2,500 multiplied by 5).

capacity factor The output of a generating plant for a specified period of time, say a year, divided by the output if the plant had operated continuously at full rated capacity for the same period.

cathodic protection A method of preventing oxidation (rusting) of exposed metal structures such as bridges by imposing between the structure and the ground a small electrical voltage that opposes the flow of electrons and that is greater than the voltage present during oxidation.

cell barrier A very thin region of static electric charge along the interface of the positive and negative layers in a photovoltaic cell. The barrier inhibits the movement of electrons from one layer to the other, so that higher-energy electrons from one side diffuse preferentially through it in one direction, creating a current and thus a voltage across the cell. Also called depletion zone, cell junction, or space charge.

cell junction The area of immediate contact between two layers (positive and negative) of a photovoltaic cell. The junction lies at the center of the cell barrier or depletion zone.

central power The generation of electricity in large power plants with distribution through a network of transmission lines (grid) for sale to a number of users. Opposite of distributed power.

combined collector A photovoltaic device or module that provides useful heat energy in addition to electricity. See *photovoltaic-thermal system.*

concentrator A photovoltaic array that includes an optical component such as a lens or focusing mirror to direct incident sunlight onto a solar cell of smaller area.

conduction band; conduction level Energy level at which electrons are not bound to (orbiting) a specific atomic nucleus but are free to wander among the atoms.

conversion efficiency (cell) The ratio of the electric energy produced by a solar cell (under full sun conditions) to the energy from sunlight incident upon the cell.

current See *electric current.*

Czochralski process Method of growing a perfect crystal of large size by slowly lifting a seed crystal from a molten bath of the material under careful conditions of cooling.

deep discharge Discharging a battery to 20 percent or less of its full charge.

dendrite A slender threadlike spike of pure crystalline material such as silicon.

depletion zone Same as cell barrier. The term derives from the fact that this microscopically thin region is depleted of charge carriers (free electrons and holes).

diffuse insolation Sunlight received indirectly as a result of scattering due to clouds, fog, haze, dust, or other substances in the atmosphere. Opposite of direct insolation.

diffusion length The mean distance a charge carrier (free electron or hole) moves before recombining with another hole or electron. Distances are short, typically several micrometers to a few hundred micrometers. Cell efficiency improves with increasing minority carrier diffusion length.

direct current (DC) Electric current in which electrons are flowing in one direction only. Opposite of alternating current (AC).

direct insolation Sunlight falling directly upon a collector. Opposite of diffuse insolation.

distributed power Generic term for any power supply located near the point where the power is used. Opposite of central power. See *stand-alone; remote site.*

distributed systems Systems that are installed at or near the location where the electricity is used, as opposed to central systems that supply electricity to grids. A residential photovoltaic system is a distributed system.

DOE U.S. Department of Energy.

donor A dopant, such as phosphorus, that supplies an additional electron to an otherwise balanced crystal structure.

dopant A chemical element added in small amounts to an otherwise pure crystal to modify its electrical properties. An n-dopant introduces more electrons than are required for the perfect structure of the crystal. A p-dopant creates electron vacancies in the crystal structure.

edge-defined film-fed growth (EFG) Process for producing a sheet of silicon. Silicon is drawn through a slit in a die or shaping guide. The cooling silicon follows the single-crystal pattern established by a seed crystal.

electric circuit Path followed by electrons from a power source (generator or battery) through an external line (including devices that use the electricity) and returning through another line to the source.

electric current A flow of electrons; electricity.

electric vehicle Automobile powered by batteries; can be charged by photovoltaic systems.

energy payback time The time required for any energy-producing system or device to produce as much useful energy as was consumed in its manufacture and construction.

EPRI The Electric Power Research Institute, Palo Alto, California; the research arm of the investor-owned utilities in the United States.

fill factor (FF) The ratio of the maximum power a photovoltaic cell can produce to the theoretical limit if both voltage and current were simultaneously at their maximums. A key characteristic in evaluating cell performance.

flat plate (module or array) An arrangement of solar cells in which the cells are exposed directly to normal incident sunlight. Opposite of concentrator.

Fresnel lens An optical device that focuses light like a magnifying glass; concentric rings are faced at slightly different angles so that light falling on any ring is focused to the same point. Fresnel lenses are flat rather than thick in the center and can be stamped out in a mold.

gallium A chemical element, atomic number 31, metallic in nature, used in making certain kinds of solar cells.

gigawatt One billion Watts. One million kilowatts. One thousand megawatts. 10^9 Watts.

grid Network of transmission lines, substations, distribution lines, and transformers used by central power systems.

heterojunction Zone of electrical contact between two dissimilar materials. See *homojunction*.

hole A vacancy where an electron would normally be in a perfect crystalline structure.

homojunction The zone of contact between the n-layer and the p-layer in a single material, the two layers having been created by doping the basic crystal with other substances. See *heterojunction*.

incident light Light that shines onto the face of a solar cell.

insolation Sunlight, direct or diffuse. Not to be confused with *insulation*. See *diffuse insolation; direct insolation*.

installed capacity See *capacity*.

inverter Device that converts DC to AC. See *rectifier*.

I-V curve A graphical presentation of the current versus the voltage from a photovoltaic cell as the load is increased from the short circuit (no load) condition to the open circuit (maximum voltage) condition. The shape of the curve characterizes cell performance.

JPL Jet Propulsion Laboratory, Pasadena, Calif.; manages the development and test efforts in photovoltaics for DOE.

kilowatt (kW) 1,000 Watts.

kilowatt hour (kWh) 1,000 Watt hours. See *Watt hour*.

kW_p See *peak kilowatt*.

load The amount of electric power being consumed at any given moment. Also, in an electrical circuit, any device or appliance that is using power. The load for a utility company varies greatly with time of day and to some extent with season. See *base load; peak load.*

majority carrier Current carriers (either free electrons or holes) that are in excess in a specific layer of a semiconductor material (electrons in the n-layer, holes in the p-layer) of a cell.

marginal cost The cost of one additional unit within a group of like units.

megawatt (MW) One million Watts; 1,000 kilowatts.

minority carrier Current carriers (either electrons or holes) that are in the minority in a specific layer of semiconductor material. It is the diffusion of minority carriers through the cell barrier that creates a voltage, and constitutes a current, in a photovoltaic device. The process becomes more efficient with increasing minority carrier diffusion length.

multiple-junction cell A photovoltaic cell containing two or more cell barriers, each of which is optimized for a particular portion of the solar spectrum to achieve greater overall efficiency in converting sunlight into electricity. See *vertical multiple-junction cell; split-spectrum cell.*

n-silicon Silicon containing a minute quantity of impurity, or dopant, such as phosphorus, which causes the crystalline structure to contain more electrons than required to exactly complete the crystal structure. There is no electrical imbalance, however. Opposite of p-silicon.

ohm A measure of resistance to the flow of an electric current.

open-circuit voltage (Voc) The voltage across a photovoltaic cell in sunlight when no current is flowing; the maximum possible voltage.

order of magnitude A factor of 10, used as a convenience in comparing large numbers.

parallel connection A method of interconnecting two or more electricity-producing devices, or power-using devices, such that the voltage produced, or required, is not increased, but the current is additive. Opposite of series connection.

payback See *energy payback time.*

peak load; peak demand The maximum load, or usage, of electrical power occurring in a given period of time, typically a day.

peak kilowatt (kW_p) The highest kilowatt output of a photovoltaic system receiving full sunlight.

peak Watt (W_p) The amount of power a photovoltaic device will produce at noon on a clear summer day with sun approximately overhead (insolation at 1,000 Watts per square meter) when the cell is faced directly toward the sun.

phosphorus (P) A chemical element, atomic number 15, used as a dopant in making n-silicon.

photoelectrochemical cell A special kind of photovoltaic cell in which the electricity produced is used immediately within the cell to produce a useful chemical product, such as hydrogen. The product material is continuously withdrawn from the cell for direct use as a fuel or as an ingredient in making other chemicals, or it may be stored and used subsequently.

photon A particle of light, which acts as an indivisible unit of energy; a quantum or corpuscle of radiant energy moving with the speed of light.

photovoltaic (PV) Pertaining to the direct conversion of light into electricity.

photovoltaic array An interconnected system of photovoltaic modules that functions as a single electricity-producing unit. The modules are assembled as a discrete structure, with common support or mounting.

photovoltaic cell A device that converts light directly into electricity. A solar photovoltaic cell, or solar cell, is designed for use in sunlight. All photovoltaic cells produce direct current (DC).

photovoltaic collector A photovoltaic module or array, which receives sunlight and converts it into electricity.

photovoltaic module A number of photovoltaic cells electrically interconnected and mounted together, usually in a sealed unit or panel of convenient size for shipping, handling, and assembling into arrays.

photovoltaic system A complete set of components for converting sunlight into electricity by the photovoltaic process, including array and balance-of-system components.

photovoltaic-thermal (PV/T) system A photovoltaic system that, in addition to converting sunlight into electricity, collects the residual heat energy and delivers both heat and electricity in usable form. Also called total energy system. See *combined collector.*

polycrystalline silicon; polysilicon Silicon that has solidified at such a rate that many small crystals (crystallites) were formed. The atoms within a single crystallite are symmetrically arrayed, whereas the crystallites are jumbled together.

power conditioner The electrical equipment used to convert power from a photovoltaic array into a form suitable for subsequent use, as in supplying a household. Loosely, a collective term for inverter, transformer, voltage regulator, meters, switches, and controls.

p-silicon Silicon containing a minute quantity of impurity, or dopant, such as boron, which provides insufficient electrons to exactly complete the crystal structure. There is no electrical imbalance, however. Opposite of n-silicon.

PV Abbreviation for photovoltaic(s).

quad (Q) One quadrillion (10^{15}) British thermal units. A commonly used measure of very large quantities of energy. The total consumption of all forms of energy in the United States in 1980 was about 78 quads.

recombination A free electron being reabsorbed into a hole.

rectifier A device that converts AC to DC. See *inverter.*

remote site Not connected to a utility grid. See *stand-alone; distributed power.*

RES Residential Experiment Station.

reserve capacity The amount of generating capacity a central power system must maintain to meet peak loads. See *spinning reserve.*

ribbon A thin sheet of crystalline or polycrystalline material, such as silicon, produced in a continuous process by withdrawal from a molten bath of the parent material.

satellite power system (SPS) Concept for providing large amounts of electricity for use on the earth from one or more satellites in geosynchronous earth orbit. A very large array of solar cells on each satellite would provide electricity, which would be converted to microwave energy and beamed to a receiving antenna on the ground. There it would be reconverted into electricity and distributed the same as any other centrally generated power, through a grid.

Schottky barrier A cell barrier established as the interface between a semiconductor, such as silicon, and a sheet of metal.

semiconductor Any material that has limited capacity for conducting an electric current. Certain semiconductors, such as silicon, gallium arsenide, and cadmium sulfide, are uniquely suited to the photovoltaic conversion process.

SERI The Solar Energy Research Institute at Golden (Denver), Colorado. Established by Congress in 1974 (Solar Energy Research, Development, and Demonstration Act) to lead the nation's solar energy research and development program; manages research effort on photovoltaics for DOE.

series connection A method of interconnecting devices that generate or use electricity so that the voltage, but not the current, is additive. Opposite of parallel connection.

short circuit current (Isc) The current flowing freely from a photovoltaic cell through an external circuit that has no load or resistance; the maximum current possible.

Siemens process A commercial method of making purified silicon.

silicon (Si) A chemical element, atomic number 14, semimetallic in nature, dark gray, an excellent semiconductor material. A common constituent of sand and quartz (as the oxide). Crystallizes in face-centered cubic lattice like diamond. See *polycrystalline silicon.*

solar cell A photovoltaic cell designed specifically for use in converting sunlight into electricity.

solar constant The strength of sunlight; 1,353 Watts per square meter in space and about 1,000 Watts per square meter at sea level.

solar thermal electric Method of producing electricity from solar energy by using focused sunlight to heat a working fluid, which in turn drives a turbogenerator.

space charge See *cell barrier; depletion zone.*

spinning reserve Utility generating capacity on-line and running at low power in excess of actual load.

split-spectrum cell A compound photovoltaic device in which sunlight is first divided into spectral regions by optical means. Each region is then directed to a different photovoltaic cell optimized for converting that portion of the spectrum into electricity. Such a device achieves significantly greater overall conversion of incident sunlight into electricity. See *multiple-junction cell.*

stand-alone An isolated photovoltaic system not connected to a grid; may or may not have storage, but most stand-alone applications require batteries or other forms of storage. See *remote site.*

synfuel, synthetic fuel Any of several fuels, usually liquid or gaseous, derived by processing such fossil sources as oil shale, tar sands, and coal.

thermal electric Electric energy derived from heat energy, usually by heating a working fluid, which drives a turbogenerator. See *solar thermal electric.*

thermophotovoltaic cell A device that concentrates sunlight on a small heat absorber made of metal or other suitable material, heating it to a high temperature. The secondary thermal radiation re-emitted by the absorber is used as the energy source for a photovoltaic cell. The cell is chosen for maximum efficiency at the wavelength of the secondary radiation.

thin film A layer of semiconductor material, for example polycrystalline silicon or gallium arsenide, typically a few hundredths of an inch or less in thickness, useful in making photovoltaic cells. Use of this material bypasses the costly steps of growing single-crystal ingots and sawing them into wafers. Depending on the material, thin films may be produced in different ways, such as withdrawing a ribbon from a molten bath, slowly cooling a molten sheet on a substrate, or spray coating.

total energy system See *photovoltaic-thermal system.*

valence state; valence level energy, bound state Energy content of an electron in orbit about an atomic nucleus.

vertical multiple-junction cell (VMJ) A compound cell made of different semiconductor materials in layers one above the other like a club sandwich. Sunlight entering the top passes through successive cell barriers, each of which converts a separate portion of the spectrum into electricity, thus achieving greater total conversion efficiency of the incident light. Also called a multiple junction cell. See *multiple-junction cell; split-spectrum cell.*

Volt, voltage (V) A measure of the force or "push" given the electrons in an electric circuit; a measure of electric potential. One volt produces one amp of current when acting against a resistance of one ohm. See *open-circuit voltage.*

wafer A thin sheet of semiconductor material made by mechanically sawing it from a single-crystal ingot.

Watt, wattage (W) A measure of electric power or amount of work done in a unit of time. One amp of current flowing at a potential of one volt produces one Watt of power.

Watt hour (Wh, Whr) A quantity of electrical energy (electricity). One Watt hour is consumed when one Watt of power is used for a period of one hour.

Watt peak (W_p) See *peak Watt.*

INDEX

Page numbers in *italic* type indicate captions.

C

H

I

J

N

O

P